Military Occupational Specialties

Change and Consolidation

Mary E. Layne

Scott Naftel

Harry J. Thie

Jennifer H. Kawata

Prepared for the
Office of the Secretary of Defense

National Security Research Division

RAND

The research described in this report was sponsored by the Office of the Secretary of Defense (OSD). The research was conducted in RAND's National Defense Research Institute, a federally funded research and development center supported by the OSD, the Joint Staff, the unified commands, and the defense agencies under Contract DASW01-95-C-0059.

Library of Congress Cataloging-in-Publication Data

Military occupational specialties : change and consolidation / Mary E. Layne . . . [et al.].
p. cm.
"Prepared for the Office of the Secretary of Defense by RAND's National Defense Research Institute."
"MR-977-OSD."
Includes bibliographical references.
ISBN 0-8330-2646-1
1. United States—Armed Forces—Occupational specialties.
2. United States—Armed Forces—Operational readiness.
I. Layne, Mary, 1958- . II. United States. Dept. of Defense. Office of the Secretary of Defense. III. National Defense Research Institute (U.S.).
UB337.M53 2001
355.4 ' 9 ' 0973—dc21 98-33559
CIP

Published 2001 by RAND
1700 Main Street, P.O. Box 2138, Santa Monica, CA 90407-2138
1200 South Hayes Street, Arlington, VA 22202-5050
RAND URL: http://www.rand.org/
To order RAND documents or to obtain additional information, contact Distribution Services: Telephone: (310) 451-7002; Fax: (310) 451-6915; Internet: order@rand.org

PREFACE

The objective of this research is to assess the extent to which skills have been eliminated and consolidated in recent years in the military Services and the resulting effects, if any, on readiness. The research supports a requirement in the report of the National Security Committee, House of Representatives, on H.R. 1119, National Defense Authorization Act for Fiscal Year 1998.

The research was conducted during a three-month period in early 1998 through the use of occupational and personnel databases available at the Defense Manpower Data Center. The work was completed and provided to the sponsor in 1998; its findings and recommendations were based on the data available at that time and have not been reviewed against present-day data. RAND is publishing the report now for archival reasons.

This research was conducted for the Assistant Secretary of Defense for Force Management Policy within the Forces and Resources Policy Center of RAND's National Defense Research Institute, a federally funded research and development center sponsored by the Office of the Secretary of Defense, the Joint Staff, the unified commands, and the defense agencies. It should be of interest to military occupational analysts and to military personnel managers.

CONTENTS

FIGURES

TABLES

SUMMARY

BACKGROUND

Over the past several years, the military Services have reduced their forces substantially, and military skill groups have been consolidated as part of that reduction. Such consolidations can benefit organizations. Workers experience greater job satisfaction because the less-restrictive job categories that result from consolidations make it easier for them to find jobs that mesh with their personal interests. Personnel management also becomes easier because the streamlining of job descriptions facilitates assignments. However, as a result of visits to military units and anecdotal reports from the field, some members of Congress became concerned that the rapid military drawdown and the accompanying skill consolidations were eroding readiness and expressed a particular concern about maintenance skills in two Services. The Office of the Secretary of Defense asked RAND to determine the extent of the consolidations and their effects on readiness.

APPROACH

To carry out this research, we surveyed the management literature, mined the databases maintained by the Defense Manpower Data Center, and interviewed experts in the Department of Defense and the military Services. We adopted a framework developed in earlier RAND work that shows that five attributes—availability, qualification, experience, stability, and motivation—provide good indicators of personnel readiness. We then analyzed the extent of skill consolidations, concentrating on the maintenance fields in the Army and the Marine Corps. Finally, we assessed the effects of consolidations on three aspects of personnel readiness for the Army: availability, qualification, and experience. The few consolidations that occurred in the Marine Corps and other constraints led us to evaluate the effect of the consolidation only on experience in that Service's maintenance occupations.

RESULTS

On the basis of the indicators selected from our framework, we found no evidence of deleterious effect of consolidation on readiness. Both the Army and the Marine Corps have detailed processes for considering and implementing consolidations of skill groups. These processes consider the rationale for the change and how it fits with overall force structure and mission. A review of the Service skill inventories shows that three Services have reduced their military occupation categories (MOCs), and these category reductions began before the drawdown. Between 1984 and 1997, the Army reduced its MOCs by about 20 percent, the Marine Corps by 13 percent, and the Air Force by 10 percent. In contrast, the Navy increased its MOCs by almost one-third, reflecting its effort to make them more weapon-system specific.

When change occurs, transitions from one MOC to another take place over about 24 months. During this period, both the loser and gainer MOCs are turbulent. Over time and especially with the aggregate data that we used, the effects of MOC disruptions smooth out. For individuals and for units, at any time within the 24-month transition window, the processes of enlarging, eliminating, and consolidating MOCs would seem less smooth because such change is not immediate and because of frictions in personnel-management processes during the period of transition.

Focusing on maintenance MOCs in the Army, we assess readiness effects based on three measures: availability, experience, and qualification. For the Marine Corps, we assess only experience.

Availability

This metric refers to level of fill of a particular MOC as a percentage of its authorization. We examined "gainer" MOCs, i.e., MOCs that received the members from the discontinued MOC. Typically, MOC fills vary. A comparison of gainer MOCs with all maintenance MOCs two years after consolidation showed that gainer MOCs enjoy higher levels of fill.

Experience

We used three indicators for experience: time in service, average pay grade, and time in grade. Again comparing gainer MOCs with all maintenance MOCs, average years of experience increased to a greater extent for the gainer MOCs (as it did over time for the other maintenance categories). The gainer MOCs became progressively more senior, as measured both by rank and average time in grade. The story in the Marine Corps somewhat parallels that in the Army: While there was variation by MOC, the experience level overall was, on average, about the same as that of maintenance MOCs. However, average pay grade was lower than all maintenance MOCs, while time in grade was higher.

Qualification

Data on qualification are more ambiguous. Few took formal courses from 6 months before consolidation to 24 months after. This could mean that considerable on-the-job training occurred, that no training was necessary, or that needed training was not given.

In sum, our indicators do not show that consolidation has adversely affected readiness. The experience and availability of personnel in consolidated MOCs are comparable with those for other maintenance MOCs. The data for qualification are less clear. However, the Services' procedures for these consolidations are clearly understood, and these procedures very likely help smooth the transitions.

ACKNOWLEDGMENTS

The Defense Manpower Data Center was helpful, as usual, in meeting our needs for data about MOCs and about the people who are assigned to them. In particular, John Fowlkes and George Nebeling receive our thanks. We are grateful to the several officials of the Army and Marine Corps who took the time to explain their processes for changing military occupational structures. We were assisted by several RAND colleagues: Jan Hanley and Laurie McDonald assisted in processing data files; Jerry Sollinger and Phyllis Gilmore assisted in report preparation; and Chip Leonard and Casey Wardynski reviewed a draft and suggested improvements.

ACRONYMS

ALMARS	All Marine Messages
ASI	Additional Skill Identifier
CG	Commanding general
CMF	Career Management Field
DMDC	Defense Manpower Data Center
DoD	Department of Defense
DoL	Department of Labor
MCCDC	Marine Corps Combat Development Command
MOC	Military Occupational Code
MOS	Military Occupational Specialty
ODB	Occupational Database
OJT	On-the-job training
PERSCOM	Total Army Personnel Command
QRMC	Quadrennial Review of Military Compensation
SQI	Special Qualification Identifier
TRADOC	Training and Doctrine Command

Chapter One

INTRODUCTION

How many and what kinds of jobs are needed to accomplish military missions? The answer rests primarily on three factors: (1) the need for forces, which is in turn driven by changes in military missions and by external events (e.g., conflict or the end of conflict); (2) organization; and (3) technology. Technology drives change in the products—military weapon systems—and in the processes—organizations. As a result, certain jobs disappear, and new ones emerge. Over long periods, significant shifts take place in how the work of the military is accomplished.[1] This report looks at the extent to which the occupational landscape of the military has changed over the past few years and whether such changes have significantly influenced readiness.

BACKGROUND

Workplace and workforce practices change over time. The 1970s were a period of emphasis on job enrichment and enlargement. Employees wanted challenging work that matched their abilities. Restrictive job categories were thought to stifle creativity and morale. The result was a move away from narrow job descriptions. Employees wanted to see results from their work, and they wanted more challenge and opportunity. For the military, for example, the Defense Resource Management Study in 1979 discussed the concept of skill broadening, which

> can help to overcome some of the compartmentalization that has accompanied recent moves toward more emphasis on task-oriented training and can permit substitution of personnel with multiple skills in places where several more people with limited skills are currently assigned. (Rice, 1979, p. 66.)

Such concepts underlie current workforce management practices that include flexible work practices, broadbanding, and use of team or group competencies.

The 1980s were a decade of worker displacement. Employees saw their jobs disappear because of economic downturns, international competition, technology, and organizational changes. Rightsizing, downsizing, restructuring, business process improvement, outsourcing, and reengineering led to substantial changes in or eliminations of jobs. Overall, the U.S. military lost some 700,000 positions, one-third of

[1]See for example, Kirby and Thie (1996).

the military workforce, between 1987 and 1996. Such massive dislocations cannot be accomplished without significant effects on workers and jobs.

In the 1990s, organizations are simultaneously enlarging and eliminating jobs. The "changing of the job" became popularized, and William Bridges captured the feelings of many in his book *Job Shift: How to Prosper in a Workplace Without Jobs* (1995). Job shift is real. In the workplace, jobs are becoming more complex and challenging at the same time they are being eliminated or significantly changed through task, organizational, and technological change.

PURPOSE OF STUDY

The Chairman of the House Committee on National Security released a report on military readiness in April 1997. In that report, which was based on visits by the committee staff to more than two dozen military installations and more than 50 military units, certain factors were cited as exacerbating the negative effects of undermanning and high operational tempo. "Of particular concern are the cutbacks in and consolidations of Military Occupational Specialties (MOSs) that have been used as a management tool to generate savings and downsize personnel." (Spence, 1997.)

In its report on the National Defense Authorization Act for Fiscal Year 1998, the National Security Committee of the House of Representatives further expressed concern about the eliminations and consolidation of MOSs that had occurred in at least two of the military Services. The committee stated that these actions, used as a management tool to accomplish personnel downsizing and generate savings, had resulted in skill shortages and imbalances, particularly in the maintenance fields. The committee wanted to know the extent of the eliminations and consolidations and their effect on readiness.

APPROACH

Our approach drew on previous research done for the Deputy Undersecretary of Defense for Readiness, on automated databases from the Defense Manpower Data Center, and on interviews with knowledgeable people in the military.

We first addressed the issue of occupational consolidation in a general way. We reviewed the literature on the theory and practice of skill consolidations in the private sector. We adapted a framework developed in a previous study (Schank et al., 1997) that suggests that five personnel attributes—availability, qualification, experience, stability, and motivation—are related to personnel readiness. *Availability* represents the numbers of people who can be assigned to positions; *qualification* measures training and capability in a duty skill; *experience* measures time spent in military service or in a skill. *Stability* deals with the issue of turbulence; stable people have been in their unit and in the same skill position for some time. *Motivation* is more intangible and has no obvious objective measure; the same set of conditions can result in different motivation levels for different people.

We analyzed the extent to which skill eliminations and consolidations have occurred. We expended most effort on enlisted personnel in maintenance fields in the Army and Marine Corps, the two Services of concern. For this study, we used the generic term Military Occupational Code (MOC) to refer to Army and Marine Corps military occupational specialties, Air Force specialty codes, and Navy enlisted classifications. We include as maintenance MOCs all specialties contained in the Department of Defense (DoD) Enlisted Occupation Codes 1 (Electronic Equipment Repairers) and 6 (Electrical/Mechanical Equipment Repairers).

Last, we used the readiness framework to assess the impact, if any, of skill eliminations and consolidations on personnel readiness. To the extent that these effects could be quantified, we did so. Otherwise, we qualitatively assessed the effects on the attributes of personnel readiness. Because the Marines experienced so few consolidations and because project resources and time were limited, the bulk of our effort was focused on the Army. Thus, we only examined the "experience" aspect of readiness for the Marine Corps.

ORGANIZATION OF THE REPORT

Chapter Two of this report reviews the literature about the theory and practice of skill consolidation in the private and military sectors. Chapter Three discusses the formal process used for determining and managing skill consolidation in the Army and Marine Corps. Chapter Four measures the extent of skill consolidations in recent years in all of the Services. Greater detail is provided for maintenance occupations in the Army and Marine Corps. Chapter Five presents quantitative evidence of the effect of skill consolidations in certain maintenance fields on the attributes of personnel readiness. Chapter Six presents our conclusions. Additional detail is provided in the appendixes.

Chapter Two

PREVIOUS RESEARCH AND PRACTICE RELATING TO SKILL CONSOLIDATION

SKILL CONSOLIDATION IN THE PRIVATE SECTOR

Skill consolidation has become common in the private sector over the past two decades. It is typically used in conjunction with other human-resource management practices, including self-managing teams, pay linked to performance, job rotation, and selection of employees based on general competencies instead of specific skills. These practices are meant to redesign how work is performed and to increase productivity, quality, flexibility, and innovation.

In some organizations, the redesign of work systems has substantially reduced the number of job classifications. For example, one corporation characterized its organizational structure as overly hierarchical, for example, with rigid job codes, inflexible work systems, an excessive focus on grades, and inhibited communication and teamwork (Donnely et al., 1992). Subsequently, this corporation consolidated 19,000 job classifications into just 200. Managers report that administration of the personnel system has been simplified, the organization is more flexible, and employees have become more productive and growth oriented.

Since job consolidation is usually only one of a package of human-resource management practices in use at any given firm, its effects are hard to quantify and evaluate. However, research has examined packages of practices as a whole and the effects they have on productivity, efficiency, and the overall financial performance of a company. In 1993, the Department of Labor (DoL) reviewed extensive research on the effectiveness of high-performance work practices (DoL, 1993).[1] Dozens of studies were reviewed, many of which were themselves large studies of hundreds of private-sector organizations. In most studies, innovative work practices were found to be positively associated with higher levels of productivity. In addition, companies using these methods tended to perform better financially, based on such measures as shareholder return, gross return on capital, and profits. An example of the research reviewed is a study of 30 comparable U.S. steel-finishing lines (Ichniowski, Shaw, and Prennushi, 1993). The lines were categorized according to their use of practices including problem-solving teams, increased training, gain-sharing plans, and reduced job classifications. The lines using these practices ran production as

[1]For other reviews of the use of innovative practices, see DoL (1994) or General Accounting Office (1991).

scheduled 98 percent of the time; lines structured more traditionally ran production as scheduled only 88 percent of the time.

The optimistic picture that prior research and DoL reports presented should not be taken to mean that a particular firm or organization would benefit simply from randomly adopting a particular policy or set of policies or that such practices would work for the military. Organizations must, of course, choose human-resource practices that work together to further the goals of the organization. Mark Huselid (1995) suggested two types of fit that should be considered when choosing management policies. *Internal fit* refers to how well a set of practices works together and interacts to promote the goals of the human-resource system. For example, policies that encourage extensive training would work well with policies that promote longer careers. *External fit* requires that human-resource management policies support the organization's competitive strategy or overall mission. Many of the practices involving teams, broad job descriptions, and pay for performance would fit an organization with a competitive strategy based on innovation.

SKILL CONSOLIDATION IN THE MILITARY

The concept of fit is useful when addressing the question of whether the military should consolidate occupational specialties or adopt other human-resource management policies that have been successful in the private sector. The 8th Quadrennial Review of Military Compensation (QRMC) strongly recommended the use of strategic human-resource management that directs peoples' activities and energies toward achieving the organization's goals (QRMC, 1997). Policies and practices should also help leaders coordinate activities by working together in a coherent system. In reviewing the long-term strategies of the military, the 8th QRMC found that many of the practices being used in the private sector should be considered appropriate for the future military, in which personnel must be highly skilled, adaptable to new challenges, and empowered to contribute in creative and innovative ways. A further suggestion was that innovative work practices may be more appropriate in different functional areas of the military.[2]

However, there is little evidence that there has been any movement toward consolidating jobs in any of the Services as part of an overall strategic plan for military human-resource management. Likewise, there has been little research on the practice of job-series consolidation and its effects in the military. However, by looking at the military and private-sector research that does exist, it is possible to develop a good sense of the benefits and possible problems associated with job consolidation. The following are some of the possible benefits:

- *Reduced costs associated with schoolhouse training.* As occupational specialties are combined, schoolhouse training may become shorter and more general, while more-specific tasks are learned through on-the-job training (OJT). Also,

[2]The idea of varying personnel policies by functional areas is further explored in Robbert et al. (1996).

eliminating training courses for specialties with low participation rates can conserve resources. (Wild and Orvis, 1993.)

- *Reduced costs associated with the personnel management infrastructure.* Reducing the total number of military occupational specialties makes the personnel system simpler to run and easier to understand. As the number of occupational specialties drops and positions become more general, it becomes easier to fill those positions.
- *Increased flexibility in assigning workload.* This increases unit efficiency because there is less chance that personnel will sit idle during lulls in the demand for their individual specialties (Gotz and Stanton, 1986). There is also the possibility of increased quality and adaptability as the work within units becomes more process oriented. Personnel with broader skills better understand how different tasks blend into a total desired output and feel more empowered to make meaningful contributions. Overall, the unit may be more efficient, adaptable, and responsive to changing environments and work demands.

Discussion of problems and risks associated with consolidation of specialties involves challenging whether the hypothesized benefits will actually be realized and, if so, their scale. In practice, consolidations of occupational specialties have often been proposed with the hope of reducing training costs through more efficient delivery of training. However, in some cases, this may simply shift training costs from the initial training schools to the units. Wild and Orvis (1993), in a study of proposed job consolidations in Army helicopter maintenance units, found that Advanced Initial Training courses for the old military occupation codes (MOC) ranged from 13 to 30 weeks of specialty training. After consolidating specialties, courses were reduced to 8 to 10 weeks of general maintenance training. But units had to devote personnel and resources to make up the difference, either by using OJT or by sending personnel to off-site courses.[3]

If training for some tasks formerly taught in initial skill training shifts to units, questions arise about whether units can provide that training. They may not have the resources or actual work assignments that cover all necessary skills included in broader, consolidated MOC. As a result, personnel serving in one unit may gain a different set of skills from personnel in the same MOC serving in another unit. The Wild and Orvis study showed that the combination of OJT and job consolidation reduced the level of training and experience in the specialized tasks that had been part of the former, separate MOC.

The result can be some degree of ad hoc specialization within the new general MOC. Personnel are no longer interchangeable. It becomes more difficult to assess the real capabilities of individuals and their depth of experience and ability in specific task areas. Personnel managers have less knowledge of personnel skills and experience, unless extensive personalization of training, experiences, and performance measurement is instituted. The need for some form of skill tracking within the new gen-

[3]It was beyond the scope of our study to address the issue of where training can be provided most efficiently, at the school or at the unit.

eral MOC would mean that the personnel management system will not be as simple or flexible as hoped. In the Army, as MOCs have decreased between 1965 and 1990, the number of Additional Skill Identifiers (ASIs), a code attached to a MOC code that indicates expertise in a specific skill, has increased proportionally, probably to continue tracking the actual skills that individuals possess.[4]

The combination of these concerns suggests that units may suffer not only monetary costs but also some reduction in unit readiness. While a unit may be adequately manned on paper, it may be less capable in practice than before consolidations took place. A concentration of personnel in a unit with only general training will significantly affect the work output of that unit. These personnel will be less productive and will require senior unit members to devote time to teaching activities, thus reducing their output as well. In the helicopter maintenance test project, some field units were designated to receive and train these new generalists, while more "front line" maintenance units were assigned only fully trained and experienced personnel.

Skill consolidation has been proposed as a way to increase unit productivity through enhanced flexibility in assigning personnel to meet changing work demands. Queuing models show there are limits on benefits from flexibility in work allocation (Wild and Orvis, 1993). Gains in productivity diminish rapidly as more MOCs are combined. By far the largest efficiency gains are made by combining two or three MOCs into one new MOC. The usefulness of broadly trained personnel is constrained by the small number of skills needed in most units.

In summary, the existing literature suggests that the following questions be addressed when evaluating particular skill consolidations:

- Will the proposed consolidations change the system in ways that support other human-resource management policies and the overall strategy and mission of the organization?
- Are there problems with quality, personnel assignment, work assignment, morale, teamwork, or training that skill consolidation can mitigate?
- Do the jobs that are to be merged have common skills and training requirements?
- Can general training that is relevant to all tasks be provided?
- Can necessary experience in all tasks be gained in all units?
- Will unit work flows and work quality benefit from cross-trained personnel?
- Can implementation be accomplished in such a way that personnel are retrained and assigned in a reasonable manner?

The next chapter explores Service policies related to job consolidation to determine if these factors are considered in recommending consolidation. We concentrate on Army and Marine Corps policies, since these are the two Services of most interest.

---

[4]These changes occurred from 1965 to 1990 (Kirin and Winkler, 1992). ASI decreased from 1985 to 1997, as did MOC.

Chapter Three

THE PROCESS OF CHANGING MILITARY OCCUPATIONAL CODE STRUCTURES

In this chapter, we review how the Army and Marine Corps go about changing occupational classification structures. Our review was conducted by examining official procedures found in directives and orders, by tracing the processes used for some previous skill consolidations, and by collecting documentation of past consolidations in the form of general announcements and schedules of assigned tasks. Individuals with responsibility for overseeing critical parts of the process were interviewed to get a sense of how decisions are made and how the system has operated.

Questions about the process for changing occupational structures tended to fall into three broad categories:

1. Do review and coordination take place at the beginning of the process to ensure fit with overall strategy and structure and with other management policies? It should be likely that bad ideas are discovered and rejected early in the process. Also, good ideas should be carefully examined and, if necessary, modified to prevent subsequent implementation problems.
2. Has a process been laid out for implementing changes once they have been approved? Changing occupational structures is a complex activity with far-reaching effects throughout the organizational system. A carefully orchestrated plan is needed to coordinate activities involving assignment, pay, incentives, training, testing, requirements determination, and coordination of databases and publications.
3. Will there be a follow-up to see how a skill consolidation or elimination has worked out over time? The system should be able to identify and make adjustments for any long-term problems that occur as the result of a change in skill classification.

ARMY

In the Army, we found a stable, well-defined process that has been in place for more than a decade. The responsibility for development, evaluation, coordination, and approval of changes in the classification structure resides with the Total Army Personnel Command (PERSCOM).[1] The approval process begins when PERSCOM receives a proposal for a MOC change from a personnel proponent. In most cases, the proponent is a Training and Doctrine Command (TRADOC) school. While developing and maintaining training courses, these schools are often the first to notice conditions that may suggest MOC changes. Such conditions include a high level of overlap in the course content for some specialties, changes in technology or equipment, dwindling enrollment, or problems filling obscure skill specialties. Other reasons for considering MOC modifications are changes in mission, doctrine, force structure, or functions. After receiving a proposal, PERSCOM is required to develop the proposal more fully; assign staff for evaluation; and notify affected Army commands, staff elements, and agencies. The proponent provides the rationale and background of proposed changes at the beginning of the process and is required to provide considerable data and analysis documenting how proposed changes will affect grade structures, authorizations, recruiting, testing, training, and other MOC changes under consideration. The proposal is reviewed and evaluated, along with comments and recommendations collected from all relevant parties. PERSCOM then makes a decision and issues notification with supporting facts and documentation.

The process for implementing approved changes is clearly laid out, along with responsibilities, tasks, and timetables.[2] PERSCOM gathers all approved changes and periodically updates data files that carry the changes to tables of organization and equipment, the Army Authorization Documents System, and the Standard Installation/Division Personnel System. Twice a year, a circular is published listing all changes to MOC structures and how implementation of these changes will specifically affect recruitment, training, reclassifications, assignment of soldiers, and manpower.[3] TRADOC is required to see that trained soldiers are available when MOC changes take effect. Soldiers currently assigned to units in the affected MOCs receive transition training, and soldiers who will graduate before changes take effect are notified. Affected soldiers are tracked to ensure they are trained and, if necessary, reassigned and reclassified. Retraining often occurs at the next permanent change of station for individual soldiers; in some cases, mobile teams of trainers may be used. Notifications are sent to unit commanders and to individual soldiers affected by changes. Notices of changes are also posted on the Internet as general announcements for career management fields and on Web sites maintained for individual skill specialties.

We did not see any requirements for follow-up evaluation to monitor the long-term effects of changes in MOC structures. This, however, does not mean that the system is unresponsive to problems. The same system that identified a need for the initial change is used to identify the need for further modifications. The system may react

---

[1]For a full description of the responsibilities and tasks associated with changing the MOC structure see Army Regulation 611-1 (Chapters 2 and 3).

[2]See Army Regulation 611-1, Chapter 3, for the detailed listing of responsibilities and the annual timetable for completion of the elements of the process.

[3]For example, Department of the Army Circular 611-96-2, Implementation of Changes to the Military Occupational Classification and Structure, 1996.

to problems in retention, training costs, or the results of occupational or Army-wide surveys.[4]

MARINE CORPS

The process for changing occupational classification structures in the Marine Corps has undergone modifications in recent years.[5] These changes reflect an effort to consolidate and coordinate all activities related to force structure development. Currently, the responsibility for developing, evaluating, and implementing changes to the MOC structure resides with the Commanding General, Marine Corps Combat Development Command (CG MCCDC) as part of the ownership of the Total Force Structure Process.[6] As owner of this process, MCCDC is required to coordinate the support and participation of functional experts and staff agencies to ensure that total force structure management procedures operate efficiently and effectively.

The placement of responsibility for evaluating and approving MOC changes within the larger processes of determining total force structure ensures consideration of organizational goals. Proposed changes to the MOC structure must fit with other elements of the organization and provide an optimal force structure that integrates people and equipment to meet specified missions in the operating forces and supporting establishment. These broad issues are examined at the very beginning of the process, before more quantitative analysis begins.

Proposals for modifying the MOC system are formally submitted to CG MCCDC. These detailed proposals must include the rationale for the change and provide supporting data on how the change would affect mission, doctrine, training, equipment, and support. Specialists in the affected occupational field and in the particular occupational specialties are involved in development and evaluation of proposals and help coordinate the involvement of other relevant staff agencies and commands. These specialists provide expertise in the classification, training, and assignment of personnel.[7] CG MCCDC then evaluates all comments, makes necessary modifications, resolves any conflicts between concerned parties, and makes a final decision.

---

[4]An example of how the system can react is MOC 33S, a technical skill specialty that involved maintenance of several computer-based systems. The level of required training for this specialty was comprehensive and, over time, had expanded to two full years. Retention became a problem because these well-trained individuals had a very high market value in the private sector. The system recognized the problem, and the MOC was split into six separate MOCs with reduced training requirements.

[5]The new directive detailing these changes has not yet been approved and published in final form, but is currently under consideration: Marine Corps Order 5311.1C, The Total Force Structure Process.

[6]See U.S. Marine Corps ALMAR 077/96 (1996) for the establishment of CG MCCDC as the owner of the Total Force Structure Process. This change was intended to ensure that force structure, manpower, training, and education plans are coordinated and complementary to each other. Management of MOC modifications is a part of this larger responsibility for establishing an optimal total force structure.

[7]For every occupational specialty there is a Military Occupational Skill Specialist who is a subject-matter expert and serves as a technical advisor in the classification, training, and assignment of personnel within a MOC. Occupational Field Managers oversee related occupational specialties and provide advice and information on proposed changes and on how changes may affect force-structure requirements and operational considerations. See Marine Corps Order 5311.1C for a full description of the duties and responsibilities of these positions.

After approval, the process of implementing the MOC modification begins.[8] CG MCCDC is responsible for the implementation process and oversees timely completion of procedures by all responsible staff agencies. Considerable coordination is required because changes must be made to MOC-related manuals, organizational tables, Marine Corps and DoD databases, and the military pay system. Skill and Service school requirements are reviewed; where necessary, changes are made to the training pipeline. The effects that MOC changes will have on personnel assignments are determined, and checks are made to ensure that qualified personnel will be assigned to the new MOC and that the changes are in compliance with grade standards.

Announcements of changes to occupational classifications are made in All Marine Messages (ALMARS). These announcements provide detailed information about the specific actions to be taken, who is affected, and the effective dates of changes. A clear listing of the responsibilities of relevant agencies is given. Details of changes in training courses or the location of training are provided for both new recruits and for personnel already serving in affected MOCs. Points of contact, with names and telephone numbers, are listed, including specialists for the affected MOCs. A review of these messages shows that an ongoing effort is made to coordinate the many activities involved in changing an occupational specialty and that additional effort is made to provide information to those who have responsibilities within the process or who are otherwise affected by a change.

As in the Army, we did not find any specific provisions for follow-up evaluation of MOC consolidation and elimination. There are, however, other means for identifying problems that may result from consolidations. Military Occupational Skill Specialists and Occupational Field Managers are continuously involved with training, assignment, and operations within skill specialties. In addition, periodic surveys of personnel are conducted to determine what work is being done in particular skill specialties and whether that work matches official job descriptions and training programs. Other periodic studies focus on groups of jobs and either validate their content or recommend change.

---

[8]See Marine Corps Order 1200.15A for a description of the MOC modification process, staff agency responsibilities, and a chronology of the process steps.

Chapter Four

EXTENT OF MOC CONSOLIDATIONS

This chapter details occupation consolidations that have occurred in each of the Services. First, it reviews the overall change in numbers of MOCs from 1984 to 1997. Next, it explores the types of changes to MOCs. Finally, it presents some observations about the nature and extent of MOC consolidations.

DATA

The data used to examine the extent of MOC consolidations come from the Occupational Database (ODB), maintained by the Defense Manpower Data Center (DMDC). The ODB is a comprehensive repository of Service occupation and related data, drawn from each Service's occupational management directives and associated data systems.

The ODB provides chronological data (to the extent possible) on gains and losses to each MOC. Losses include obsolescence, complete transfer into another occupation, and transferring of a subset of the occupation's duties into another MOC.[1] Gains to a MOC include creation, integration of portions of another MOC's duties, and consolidation of another occupation into the MOC. Table 4.1 gives an example of the richness of information provided in the ODB for Army maintenance MOC 35M (Radar Repairer).

OVERALL TRENDS IN TOTAL MOC NUMBERS

Figure 4.1 presents, for each military Service, the number of MOCs from 1984 to 1997.[2] This long period enables us to determine if there was a downward trend that predated the Armed Forces drawdown (which began in 1988 but had its largest reductions beginning in 1990). And, indeed, there was a downward trend in the total number of MOCs for every Service except the Navy before the official drawdown period.

[1]For the Army, transfers may also be to Additional Skill Identifier (ASI) or a Special Qualification Identifier (SQI). An ASI identifies specialized skills, qualifications, and requirements that are closely related to and are in addition to those inherent in the MOS. ASIs are authorized for use with designated MOSs and are listed in each specification for these MOSs. SQIs are authorized for use with any MOS to identify special requirements. During the period covered by our study, the Army reduced the number of ASIs and SQIs. In particular, since 1985, ASIs have been reduced by 20 percent. In 1997, there were 121 ASIs and 19 SQIs.

[2]Data were culled from the Occupational Conversion Index (DoD, 1984, 1989, 1993, and 1997).

Table 4.1

History for Army MOC 35M (Radar Repairer)

Date	Gain/Loss
June 1966	35M created
September 1993	SQI (below) authorized for use with 35M
	I (Installer)
April 1994	SQI (below) authorized for use with 35M
	T (Technical Inspector)
	Y (Pathfinder)
July 1994	SQI (below) authorized for use with 35M
	C (Nuclear, Biological, and Chemical)
	K (Pathfinder)
	7 (Special Operations, Aviation)
	R (Research, Development, Test, and Evaluation)
April 1995	Some portion of ASI 8A (Identification of Friend or Foe, IFF, Repairer) duties were integrated into 35M
April 1995	The following MOC were consolidated into 35M
	13R (Field Artillery Firefinder Radar Operator)
	39C (Target Acquisition/Surveillance Radar Repairer)

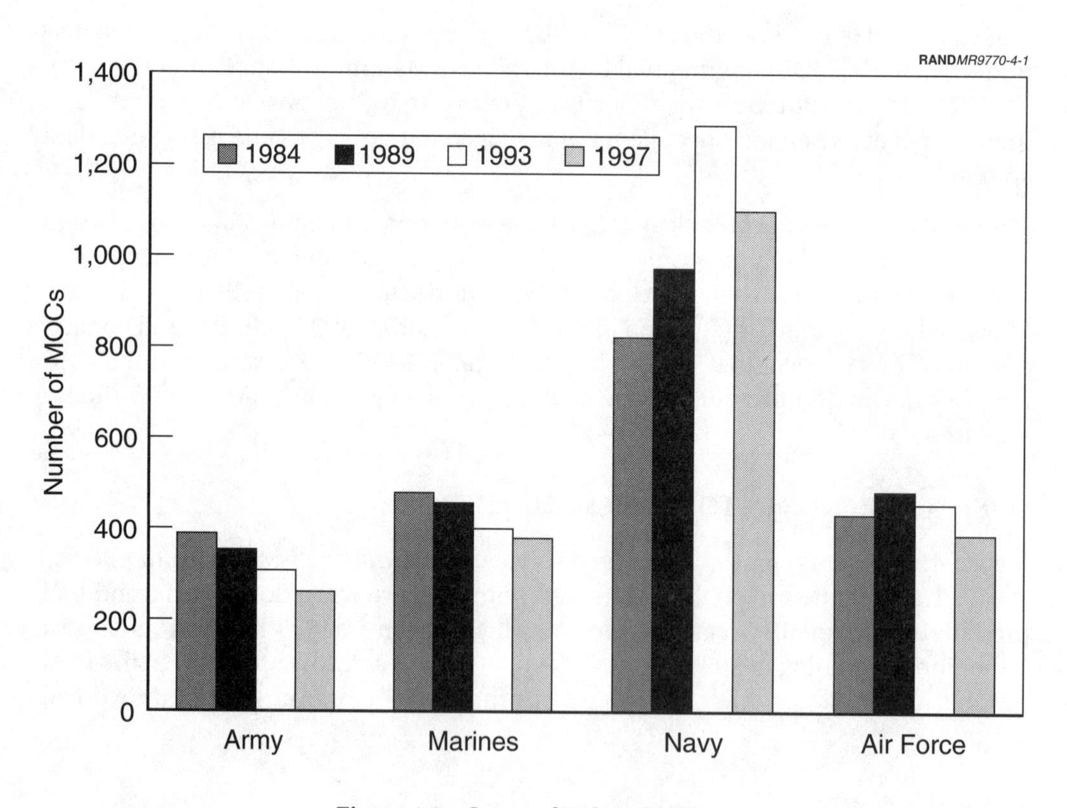

Figure 4.1—Count of Enlisted MOCs

Between 1984 and 1997, the number of Army MOCs decreased by 20 percent, and Marine Corps MOCs decreased by 13 percent. In contrast, the Navy MOCs increased by 34 percent during the same period. In 1993, the Navy expanded skills to make each MOC more weapon-system specific, hence the jump in the number of MOCs. Since then, however, the number of Navy MOCs has decreased by 15 percent.[3] Finally, Air Force MOCs declined by 10 percent over the period.

One reason for the decreasing numbers of MOCs is elimination due to the final obsolescence of a weapon system. For example, during this period, the Army saw many changes in armor, artillery, air defense, and aviation systems. Another reason is MOC consolidation, which can take two forms: (1) transferring one MOC to another or to many other existing MOCs, and (2) transferring one or many MOCs into a single new MOC. Table 4.2 examines MOC deletions and consolidations in a representative year (1995) for the Army. Either type of MOC change can cause turbulence. The first column ("Loser") is the MOC that was deleted or consolidated; the second column ("Gainer") is the MOC expanded (if applicable); the third column ("Type") indicates whether the transaction is a consolidation or deletion; and the fourth and fifth columns give the titles associated with the losing and gaining MOCs, respectively. In 1995, 15 percent of changes were deletions.[4]

An example of the first type of consolidation is MOC 31M (in the "Loser" column in Table 4.2), which was divided into two MOCs. Its duties, functions, positions, and personnel were transferred to MOC 31R, while all positions and selected personnel associated with ASI 7A (Satellite Communications Terminal AN/TSC-85/93 Operator/Maintainer) were transferred to MOC 31S. An example of the second kind of consolidation is MOC 29N (in the "Loser" column in Table 4.2); its duties, functions, positions, and personnel were transferred to the new MOC 35N.[5] Examples of MOCs deleted because of weapon-system obsolescence are HAWK Missile MOC 16D, 16E, 24C, 24G, and 24R. In this case, people with these skills were transferred to other MOCs.

FREQUENCY AND TYPE OF MOC CHANGES

This section examines the kinds of changes to MOC classifications for the Army, Marine Corps, and Air Force (data for the Navy were unavailable). We created two categories of change type: duty (task) transfers and MOC consolidations. We consider duty transfers less disruptive to the MOC (e.g., transfers to or from an ASI, SQI, or another MOC), while consolidations are viewed as more disruptive to the receiving MOC because it must absorb personnel with perhaps unrelated skills. Further, incoming personnel have to be trained in the new skills.

---

[3]We have not pursued the Navy change in depth because we were charged with investigating the consolidations and the associated effects on readiness. The Navy has had the fewest consolidations over time and is thus least likely to have had readiness problems resulting from consolidations. The large number of MOCs in the Navy and the increases through 1993 are partly due to a greater variety of weapon systems and technology and partly due to a management perspective that favored very specific and narrow MOCs. Some of the reduction since 1993 may be due to a realization by personnel managers that job categories had become too specific, causing problems in work efficiency and personnel assignment.

[4]Not all deletions are due to weapon-system obsolescence. Moreover, consolidations often presage deletions due to weapon-system obsolescence.

[5]MOS 35N was established to identify positions requiring soldiers qualified to perform DS/GS maintenance on switchboards, telephones, and associated wire instruments.

Table 4.2

Army MOC Deletions and Consolidations, 1995

Loser	Gainer	Type	Loser Title	Gainer Title
00E	79R	Consolidation	Recruiter	Recruiter
00R	79R	Consolidation	Recruiter/Retention NCO	Recruiter
00R	79S	Consolidation	Recruiter/Retention NCO	Career Counselor
01H	—	Deletion	Biological Sciences Assistant	
13R	35M	Consolidation	Field Artillery (FA) Firefinder Radar Operator	Radar Repairer
16D	—	Deletion	HAWK Missile Crewmember	
16E	—	Deletion	HAWK Fire Control Crewmember	
24C	—	Deletion	HAWK Firing Section Mechanic	
24G	—	Deletion	HAWK Information Coordination Central Mechanic	
24R	—	Deletion	HAWK Master Mechanic	
27B	27E	Consolidation	Land Combat Support System (LCSS) Test Specialist	Land Combat Electronic Missile System Repairer
27B	27M	Consolidation	Land Combat Support System (LCSS) Test Specialist	Multiple Launch Rocket System (MLRS) Repairer
27B	35B	Consolidation	Land Combat Support System (LCSS) Test Specialist	Land Combat Support System (LCSS) Test Specialist
27B	35Y	Consolidation	Land Combat Support System (LCSS) Test Specialist	Integrated Test Equipment Family Operator/Maintainer
27F	—	Deletion	VULCAN Repairer	
27J	27H	Consolidation	HAWK Field Maintenance Equipment/Pulse Acquisition Radar Repairer	HAWK Firing Section Repairer
29E	35E	Consolidation	Radio Repairer	Radio and Communications Security (COMSEC) Repairer
29J	35J	Consolidation	Telecommunications Terminal Device Repairer	Telecommunication Terminal Device Repairer
29N	35N	Consolidation	Switching Central Repairer	Wire Systems Equipment Repairer
29S	35E	Consolidation	Communications Security Equipment Repairer	Radio and Communications Security (COMSEC) Repairer
29S	35W	Consolidation	Communications Security Equipment Repairer	Electronics Maintenance Chief
29W	35W	Consolidation	Communications Electronics Maintenance Chief	Electronics Maintenance Chief
29Z	31Z	Consolidation	Electronics Maintenance Chief	Senior Signal Sergeant
29Z	35W	Consolidation	Electronics Maintenance Chief	Electronics Maintenance Chief
29Z	35Z	Consolidation	Electronics Maintenance Chief	Senior Electronics Maintenance Chief
31D	31R	Consolidation	Mobile Subscriber Equipment (MSE) Transmission System Operator	Multichannel Transmission Systems Operator-Maintainer
31M	31R	Consolidation	Multichannel Transmission Systems Operator	Multichannel Transmission Systems Operator-Maintainer
31M	31S	Consolidation	Multichannel Transmission Systems Operator	Satellite Communications Systems Operator-Maintainer

Table 4.2—Continued

Loser	Gainer	Type	Loser Title	Gainer Title
31Y	31S	Consolidation	Telecommunications Systems Supervisor	Satellite Communications Systems Operator-Maintainer
31Y	31W	Consolidation	Telecommunications Systems Supervisor	Telecommunications Operations Chief
36L	31F	Consolidation	Transportable Automatic Switching Systems Operator/Maintainer	Network Switching Systems Operator-Maintainer
36L	74G	Consolidation	Transportable Automatic Switching Systems Operator/Maintainer	Telecommunications Computer Operator-Maintainer
39C	35C	Consolidation	Target Acquisition/Surveillance Radar Repairer	Surveillance Radar Repairer
39C	35M	Consolidation	Target Acquisition/Surveillance Radar Repairer	Radar Repairer
39E	35F	Consolidation	Special Electronic Devices Repairer	Special Electronic Devices Repairer
39G	74G	Consolidation	Automated Communications Computer Systems Repairer	Telecommunications Computer Operator-Maintainer
43E	92R	Consolidation	Parachute Rigger	Quartermaster
57F	92M	Consolidation	Mortuary Affairs Specialist	Quartermaster
74D	74B	Consolidation	Information Systems Operator	Information Systems Operator-Analyst
74F	74B	Consolidation	Software Analyst	Information Systems Operator-Analyst
79D	79S	Consolidation	Reenlistment NCO	Career Counselor
83E	81L	Consolidation	Photo and Layout Specialist	Lithographer
83F	81L	Consolidation	Printing and Bindery Specialist	Lithographer
88Y	88K	Consolidation	Marine Senior Sergeant	Watercraft Operator
88Y	88Z	Consolidation	Marine Senior Sergeant	Transportation Senior Sergeant
91N	91B	Consolidation	Cardiac Specialist	Medical Specialist
94B	92G	Consolidation	Food Service Specialist	Quartermaster

Figures 4.2 and 4.3 detail the level of MOC activity in the Army from 1984 through 1997 for all MOCs and for the maintenance MOCs. While there were consolidations between 1991 and 1996, the bulk of activity was attributable to less-disruptive duty transfers. In earlier years, the bulk of the activity was of the consolidation type. Many of the more disruptive consolidations preceded the drawdown and cannot be attributed to it.

Similarly, Figures 4.4 and 4.5 detail the level of MOC activity in the Marine Corps for all MOCs and for the maintenance MOCs. Here also, we observe that, in the later years (1994–1996), changes were largely of the duty-transfer variety, while changes in the earlier years (1989–1993) were mostly consolidations. There was one consolidation in 1997, and there were no classification changes in 1992. (The ODB has no Marine data from before 1989, and there was no MOC activity between 1990 and 1993.)

The Air Force presents a somewhat different trend in the later years (Figures 4.6 and 4.7), with more consolidations than duty transfers in 1992 and 1995. As with the Army and Marines, changes in the earlier years were overwhelmingly consolidations. We excluded 1993 from this analysis because the Air Force completely changed its MOC designations that year, and there were no maintenance MOC classification changes in 1991.

Figures 4.8 and 4.9 provide a closer look at the makeup of consolidations for the Army and Marine Corps, respectively, by examining the proportion of consolidations that occurred in maintenance and nonmaintenance MOCs. Maintenance MOCs in the Army represented a large proportion of the consolidations—over half in many of the years. Of the 316 consolidations that took place from 1984 to 1997, maintenance MOCs accounted for 54 percent. The Marines display an interesting pattern: In some years, maintenance MOCs accounted for most (or all) of the consolidations; in other years, none occurred. For 1989 to 1997, maintenance MOCs accounted for 41 percent of total consolidations (54).

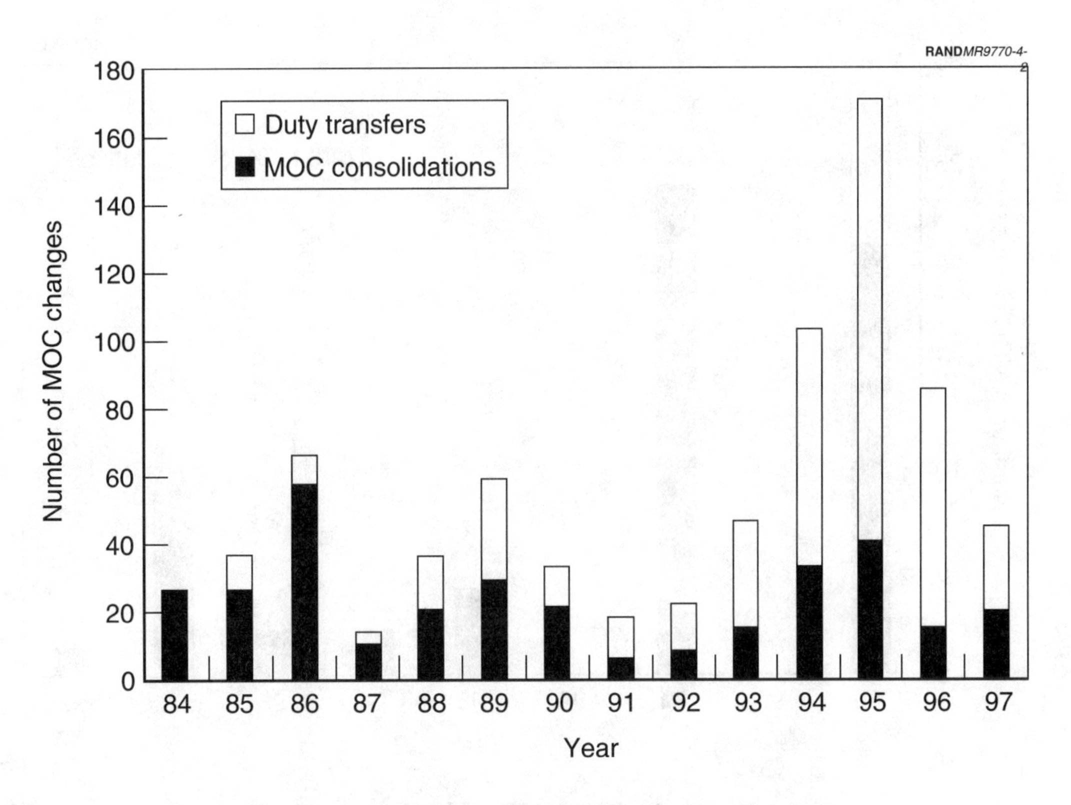

Figure 4.2—Army Enlisted MOC Classification Changes

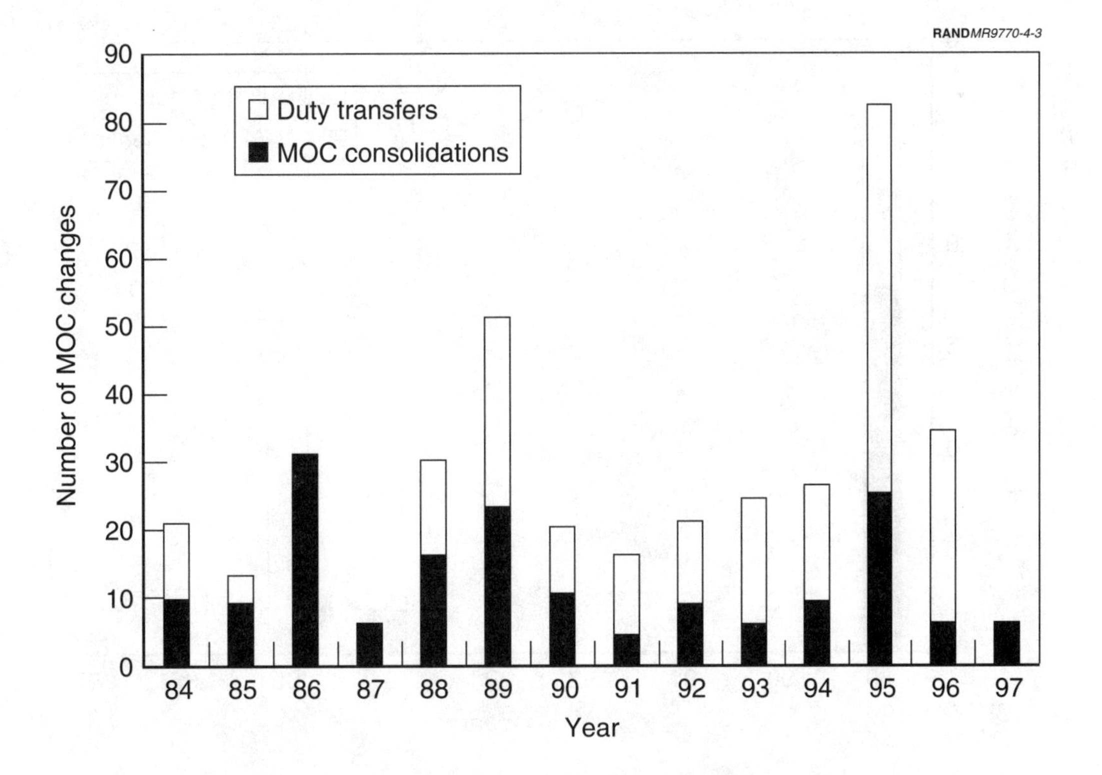

Figure 4.3—Army Enlisted Maintenance MOC Classification Changes

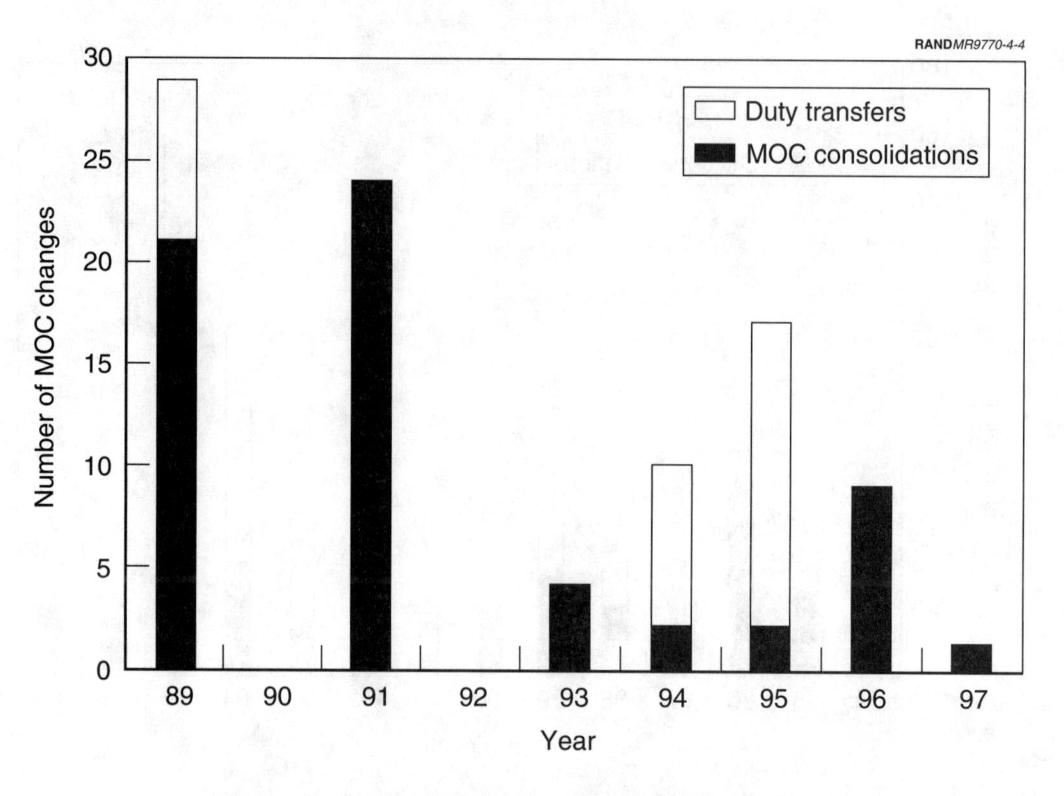

Figure 4.4—Marine Enlisted MOC Classification Changes

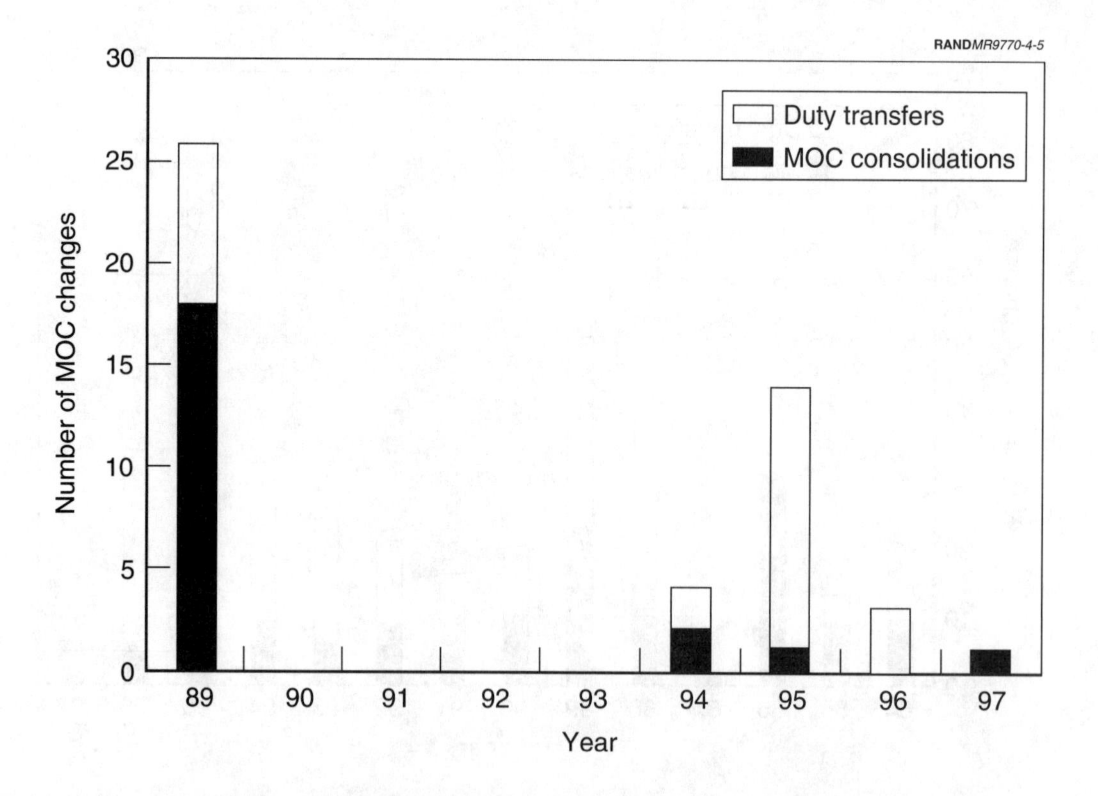

Figure 4.5—Marine Enlisted Maintenance MOC Changes

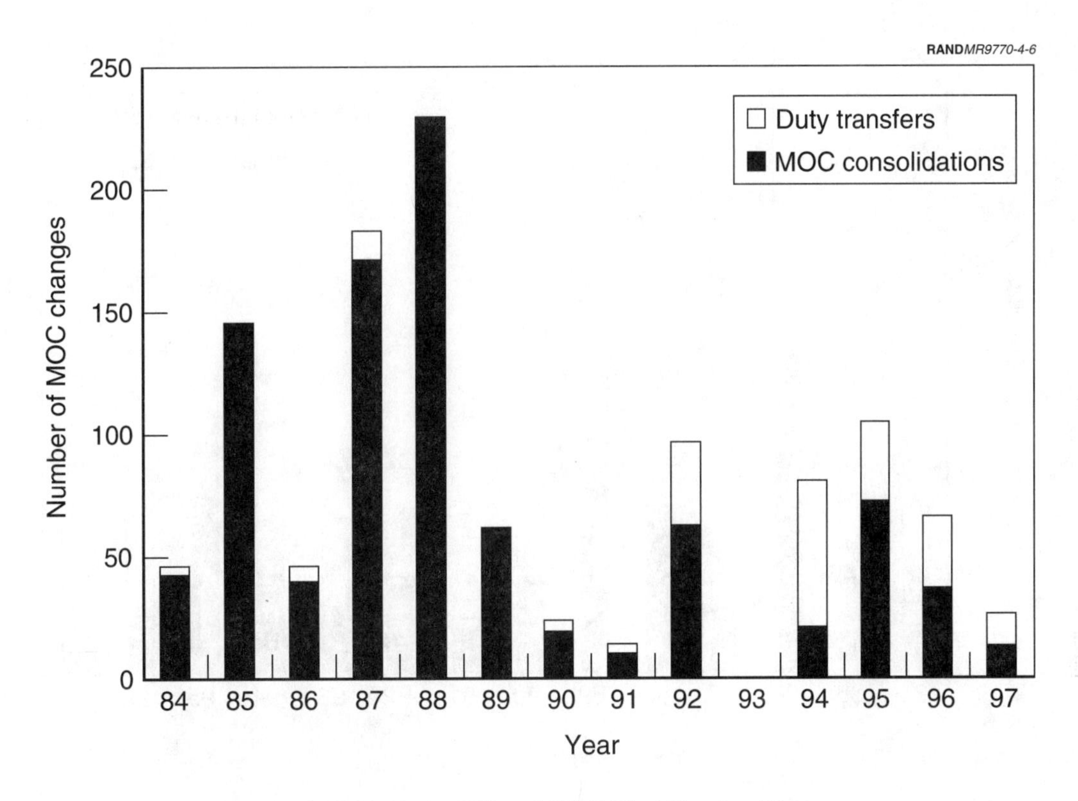

Figure 4.6—Air Force Enlisted MOC Classification Changes

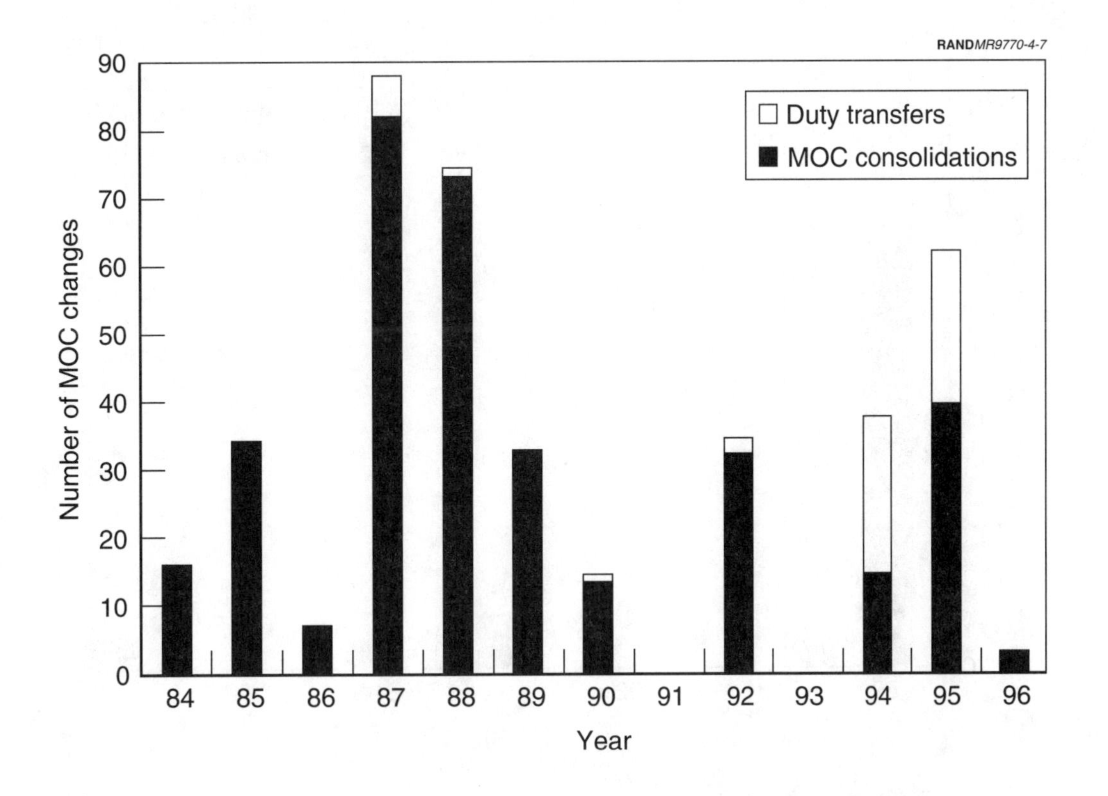

Figure 4.7—Air Force Enlisted Maintenance MOC Classification Changes

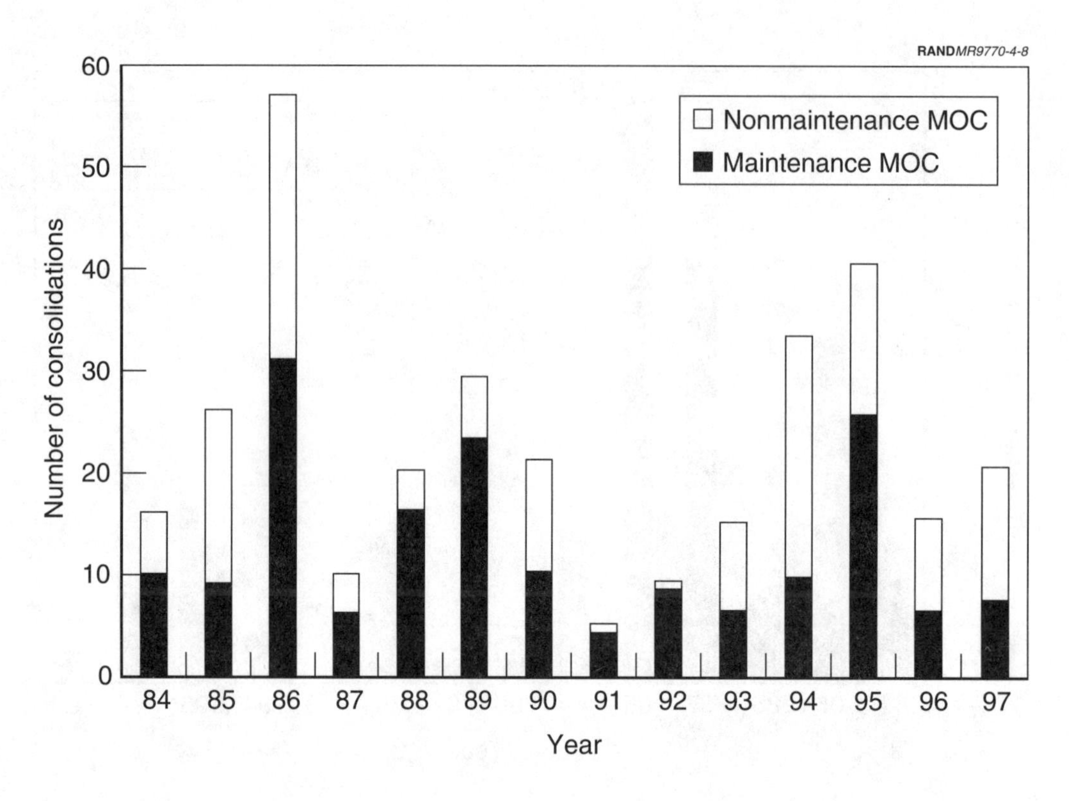

Figure 4.8—Composition of Army Consolidations

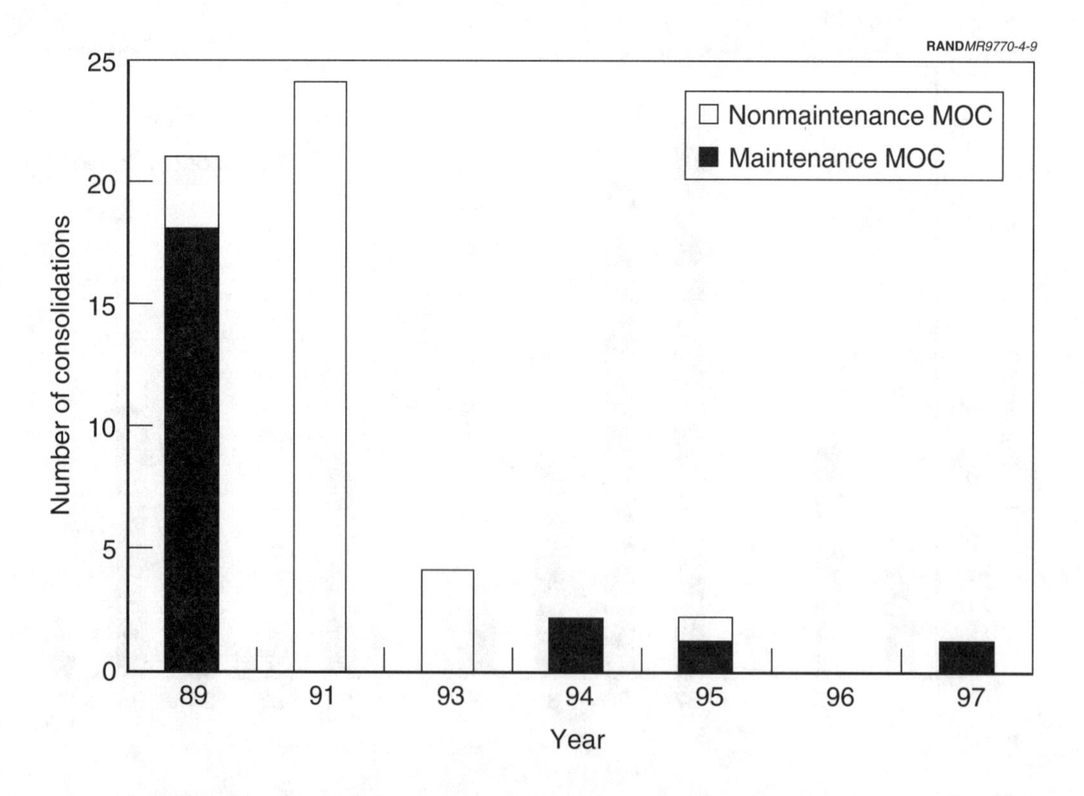

Figure 4.9—Composition of Marine Corps Consolidations

Chapter Five

EFFECT OF MAINTENANCE MOC CONSOLIDATION ON READINESS

This chapter details the effects of maintenance MOC consolidations on readiness for the Army and—on a more limited basis—the Marine Corps. (Appendixes A and B detail all consolidations in maintenance MOCs from 1984 to 1997 for these two Services.) First, it discusses the framework we used to measure readiness and the specific ways in which the framework was implemented for this study. Next, it examines, in greater detail, active-duty enlisted Army and Marine Corps maintenance MOCs before and after consolidation. Our primary focus is the Army because the Marines saw so few consolidations. The chapter concludes with observations on the effects of MOC consolidations on readiness.

READINESS FRAMEWORK

Previous research (Schank et al., 1997) has suggested that personnel readiness has five attributes:

1. *Availability*—percentage of required or authorized personnel available
2. *Experience*—percentage fill of senior grades and longevity of the force
3. *Qualification*—percentage of personnel qualified in their duty skills
4. *Motivation*—the emotional well-being of personnel
5. *Stability*—the length of time personnel have been assigned to a unit and skill.

Ideally, we would use all five of the attributes to measure the impact of MOC consolidations on readiness. In fact, we applied only three of the five factors. *Motivation* was dropped because there is no clear quantitative way to measure emotional well-being. Our time and resources were limited, so *stability* was also considered outside the scope of this study because it required data at the unit level.

The remaining factors were measured as follows:

- *Availability*—compared actual inventory with authorizations
- *Experience*—calculated time in service, pay grade, and time in grade
- *Qualification*—examined training histories for personnel in consolidated maintenance MOCs.

We assessed these measures at different stages of a MOC consolidation: 6 months prior, the consolidation date itself, 6 months later, 1 year later, 18 months later, and 2 years later. These time points were selected to capture the critical times of a consolidation: before, during (which we hypothesized would be a tumultuous period), and after.

DATA

The analysis required longitudinal data so that readiness measures were contemporaneous with the consolidations. We used several data sources in concert with the ODB. The primary data source was the DMDC Active Duty Master File, which provides an inventory, by Service, for all individuals at distinct times. These data are available quarterly for 1984 to 1991 and monthly for 1992 to 1995.[1] For the Army, we obtained an additional year of data (1996) from the Enlisted Master File.

Army training data, used to measure *qualification*, were obtained from the Army Training Requirements and Resources System. This system supports the planning, budgeting, management, and program execution phases of the training process and provides an ongoing evaluation of these phases. Records include dates, titles, and duration of courses. Training data prior to 1993 were considerably less informative. For these years, we were able to determine only if a particular person received training, not the exact type or duration. Authorization data for the Army, used to measure *availability*, were culled from the Personnel Management Authorization Document. This document is updated annually, and issues were available for 1988 through 1996. For all tables referred to in this chapter, "Loser" is the nomenclature for the MOC that was deleted and consolidated into the "Gainer" MOC; "Date" is the official date of consolidation.

MOC CONSOLIDATIONS TAKE TWO YEARS TO IMPLEMENT

MOC changes are not instantaneous. In fact, our review of the data suggests that a period of turbulence is introduced that lasts for up to two years. After that period, the system adjusts to the new MOC structure. During this transition from one state to another, leeway exists for more errors in personnel management, such as misassignments, to occur. Moreover, particular units involved in the transition bear the brunt of the frictions as the change takes place. At the end of the transition, the new state might be qualitatively better if the expected benefits materialize. However, if, before the MOC change, a military Service does not have enough authorizations overall compared to workload or fewer personnel than authorizations, a similar situation will exist after the MOC transition. A MOC change cannot be expected to resolve such preexisting conditions in the overall system as too many or too few authorizations or too few personnel.

Army

Table 5.1 details counts of enlisted personnel for all loser MOCs, summarized by the official date of consolidation. The counts are reported at various times: 6 months

---

[1]While DMDC has data for 1996 and 1997, limited time and resources prevented us from obtaining and using it.

prior to the consolidation date, the consolidation date, and at 6-month intervals for the next 2 years. Over time, all loser MOCs decrease in strength. Before the official consolidation date, strength in the MOC begins to fall. After the official consolidation date, it takes several years for the old MOC to fade from the Active Duty Master File; most of the strength in the loser MOC had left the inventory within 2 years.

Figure 5.1 shows the proportion of personnel remaining in loser MOCs at 6-month increments after the date of consolidation. As shown in this figure, the consolidations are typically implemented over time, generally with less than 10 percent of personnel remaining after 2 years.

Table 5.2 gives corresponding counts of enlisted personnel in "gainer" MOC. Over time, a gainer MOC might increase significantly in strength, stay relatively constant, or even decline in strength. This is attributable to the downsizing occurring in the Army overall or in maintenance MOCs as a group, which might cause authorizations to change. The table shows that turbulence in strength—down as well as up—occurs over a 2-year period for MOCs that ostensibly should be gaining in strength. Across the 10 years of data in Table 5.2, 27 percent of individual MOCs decreased by 20 percent or more in strength between the effective data and 2 years later, while 38 percent increased in strength by 20 percent or more. Of the latter group, nearly 60 percent doubled in strength over the 2-year period. Figure 5.2 shows the percentage change in gainer MOCs for selected time periods.

Table 5.1

Numbers of Enlisted Personnel in Loser MOC Before and After the Transition (t), in 6-month Increments, Army

Date	t—6	t	t+6	t+12	t+18	t+24
8603	—	—	—	—	—	1,274
8604	—	—	—	—	—	122
8610	—	—	—	—	1,482	683
8704	—	—	—	189	73	35
8809	1,906	1,806	1,434	713	266	122
8810	2,164	2,029	1,744	961	460	222
8903	1,586	1,614	1,505	652	327	232
8909	1,281	1,278	1,136	740	422	216
8910	257	231	159	90	67	24
9003	12	2	1	—	—	—
9009	52	43	22	13	8	2
9010	2,136	1,993	1,609	849	46	24
9104	993	948	684	62	22	16
9107	20	32	12	8	1	—
9203	208	149	79	35	17	5
9210	1,546	1,031	571	298	98	54
9304	10,432	9,382	5,090	2,059	1,049	493
9310	104	93	50	16	3	2
9404	3,816	3,601	2,151	989	440	212
9504	14,719	13,607	8,317	3,871	1,780	—
9604	1,191	1,097	534	—	—	—

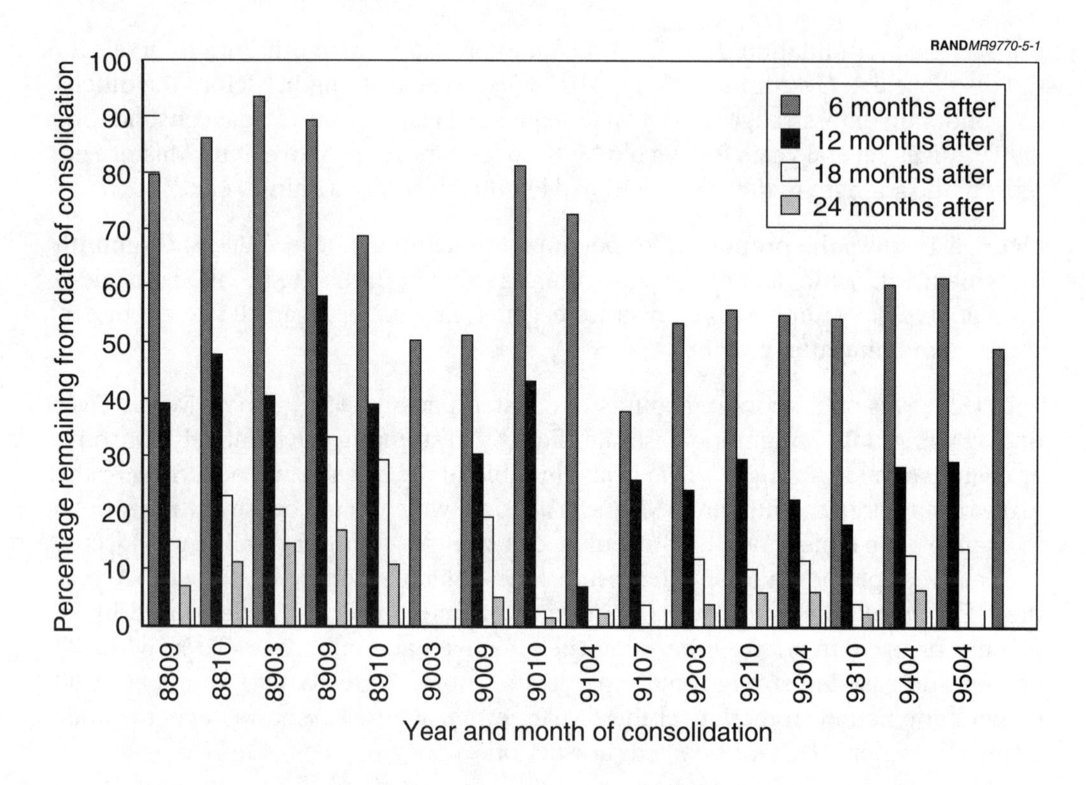

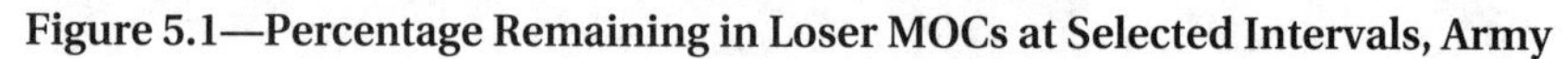
Figure 5.1—Percentage Remaining in Loser MOCs at Selected Intervals, Army

Table 5.2

Numbers of Enlisted Personnel in Gainer MOCs Before and After the Transition (t), in 6-month Increments, Army

Date	t-6	t	t+6	t+12	t+18	t+24
8603	—	—	—	—	—	3,735
8604	—	—	—	—	—	1,267
8610	—	—	—	—	8,645	9,438
8704	—	—	—	540	528	672
8809	924	1,063	1,316	2,044	2,205	2,087
8810	1,801	2,006	2,095	2,798	3,171	3,470
8903	1,190	1,154	1,319	2,030	2,336	2,427
8909	12,586	12,703	12,376	12,269	12,598	11,685
8910	3,194	3,410	3,365	3,385	3,285	2,976
9003	223	244	222	217	191	168
9009	716	755	786	832	873	821
9010	1,235	1,345	1,593	1,984	2,670	2,262
9104	1,756	1,977	2,088	2,697	2,608	
9203	1,055	986	876	856	828	792
9210	4,733	5,374	5,580	5,500	5,805	5,813
9304	3,746	4,039	7,738	10,322	10,872	11,042
9310	326	346	370	361	350	311
9404	458	569	1,860	2,950	3,300	3,244
9504	6,310	7,354	11,198	13,903	14,129	—
9604	2,648	2,398	2,524	—	—	—
9604	1,191	1,097	534	—	—	—

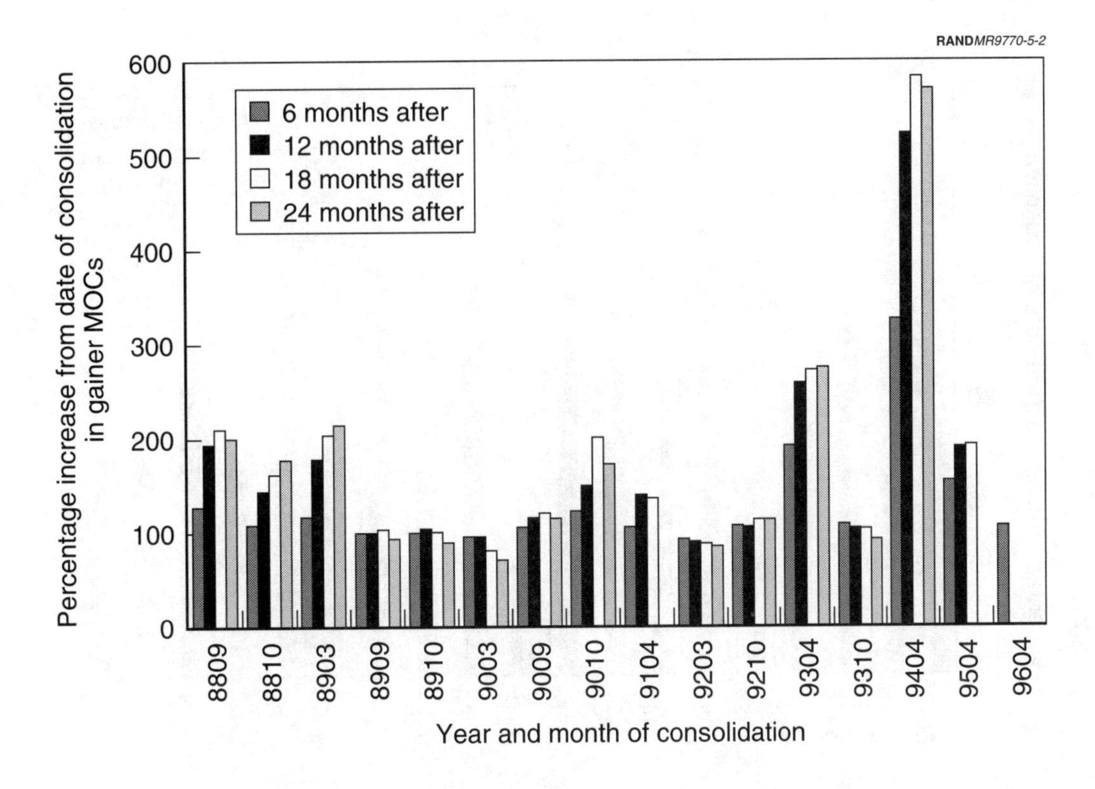

Figure 5.2—Percentage Increase in Personnel Assigned to Gainer MOCs at Selected Intervals, Army

Marine Corps

Appendix B lists consolidations in maintenance MOCs from 1988 to 1997. Figure 5.3 details the proportion of personnel remaining in loser MOCs at 6-month increments after consolidation. Changes are implemented over time; at the end of 2 years, an average of 28 percent of personnel remained in the "loser" MOC (versus 10 percent in the Army "loser" MOC).

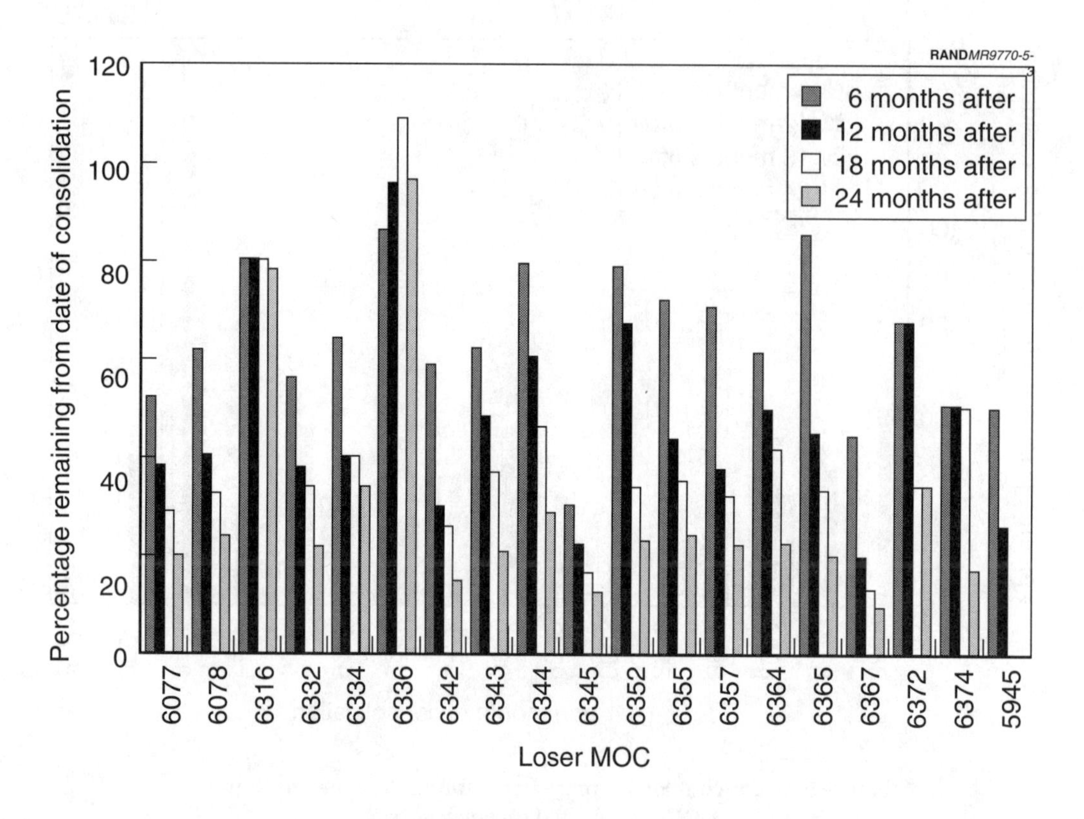

Figure 5.3—Percentage Remaining in Loser MOCs at Selected Intervals, Marine Corps

EFFECTS ON ARMY READINESS

Availability

Were personnel available to fill authorizations after consolidations? We examined the authorized and actual strengths at various periods after consolidation for gainer MOCs from FY88 to the present. Fill percentages varied, as they do for all MOCs in the Army. Table 5.3 presents the data for gainer MOCs.

Figure 5.4 arrays the fill percentages for gainer MOCs from low to high. Most MOCs were within a band of 80 to 120 percent of authorizations.

Figure 5.5 shows fill percentages two years after consolidation for gainer MOCs compared to all Army MOCs and all maintenance MOCs. On average, gainer MOCs have higher fill percentages. It is likely that MOCs that have undergone consolidations have had authorizations carefully scrutinized and personnel management plans carefully reviewed, because that is part of the change process. Moreover, these MOCs are likely to receive special attention during implementation from recruiting, training, distribution, and assignment managers.

Table 5.3

Fill of Gainer MOC 24 Months After Date of Consolidation

Gainer MOC	Date	Authorized	Actual	Fill (Percent)
27J	8809	166	147	89
27V	8809	57	24	42
31N	8809	1,845	1,916	104
24H	8810	118	107	91
24K	8810	185	192	104
27H	8810	262	223	85
39V	8810	88	84	95
68J	8810	1,609	1,764	110
68N	8810	1,124	885	79
68P	8810	241	124	51
29J	8903	977	1,055	108
39G	8903	176	195	111
68L	8903	494	371	75
68Q	8903	286	253	88
68R	8903	763	450	59
67G	8909	13	16	123
67H	8909	671	251	37
67N	8909	2,183	1,774	81
67R	8909	1,218	1,428	117
67S	8909	322	361	112
67T	8909	3,095	2,923	94
67U	8909	1,990	1,981	100
67V	8909	1,501	1,443	96
67Y	8909	1,393	1,503	108
24H	8910	64	64	100
24K	8910	89	104	117
29J	8910	1,064	876	82
68J	8910	801	1,383	173
25L	9003	119	139	117
29W	9009	665	609	92
29Z	9009	251	212	84
29S	9010	1,470	1,400	95
29Z	9010	202	193	96
33Y	9010	451	568	126
29V	9104	1,046	1,268	121
29Y	9104	1,127	1,237	110
29J	9203	746	709	95
31F	9210	4,307	3,693	86
45B	9210	491	490	100
45G	9210	374	344	92
45K	9210	1,143	1,110	97
31U	9304	8,072	7,856	97
55B	9304	2,673	2,664	100
33R	9310	226	217	96
27Z	9404	51	59	116
31P	9404	1,507	1,521	101
31S	9404	1,714	1,449	85
Total		49,430	47,635	96

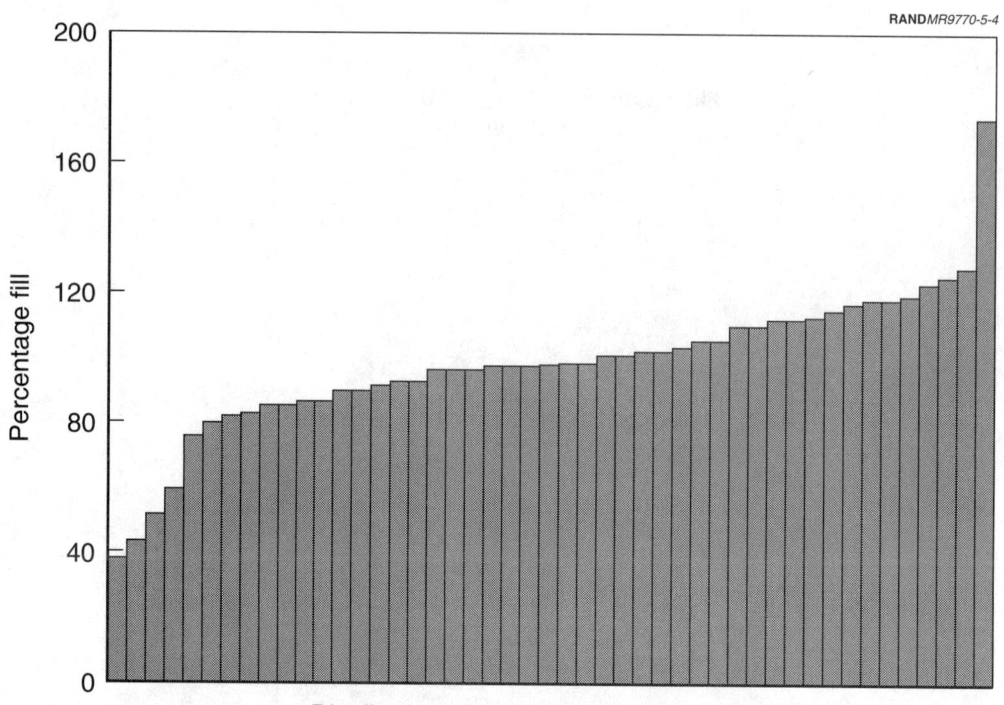

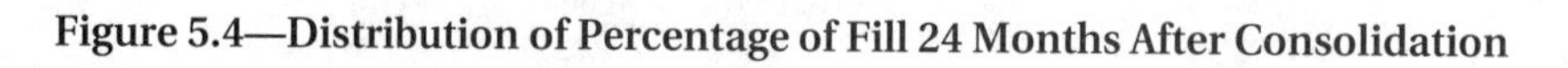

Figure 5.4—Distribution of Percentage of Fill 24 Months After Consolidation

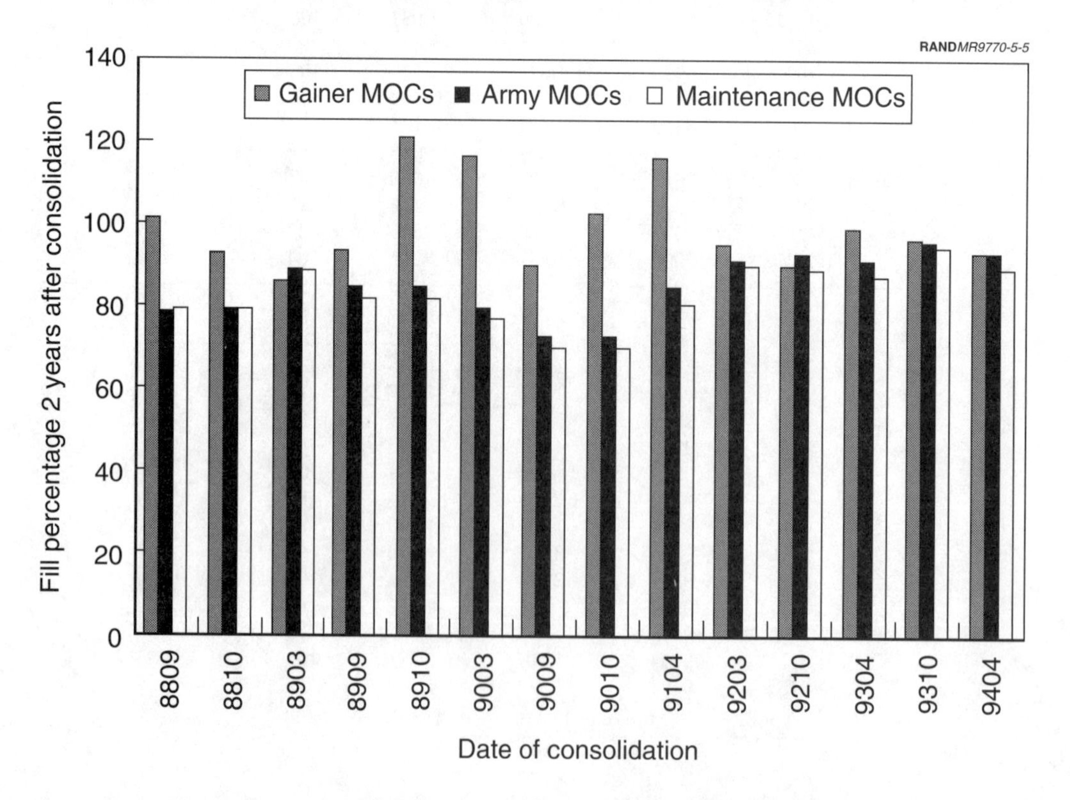

Figure 5.5—Percentage of Fill 24 Months After Consolidation

Experience

Table 5.4 presents average years of service for individuals in gainer MOCs, at various points during the consolidation period. Also shown in the table are average years of service in all maintenance MOCs, and for the total Army, contemporaneous with the consolidation. For the bulk of the consolidations, average years of experience increased over the consolidation period. Similarly, average years of experience also rose for maintenance occupations as a whole and for the total Army.

Figure 5.6 shows years of service at various times for all gainer MOCs compared to all Army MOCs and all maintenance MOCs. At consolidation, personnel in gainer MOCs are less experienced on average than all Army or all maintenance personnel. Over time, however, personnel in gainer MOCs become more experienced and exceed counterpart averages after 18 months.

Table 5.5 shows average pay grade as a measure of seniority. A force is considered more senior if more personnel are in the E5 to E9 range, or in the "Top 5." Thus, higher average grades would represent a force that is senior. Lower average grades would represent a more junior force; in particular, anything below 4 would be considered entry level. In most of the consolidations, we see that gainer MOCs become progressively more senior.

Table 5.4

Average Years of Service in Gainer, Army Enlisted, and Maintenance MOCs at Selected Times

	Gainer MOCs			Army MOCs			Maintenance MOCs		
Date (t)	t	t+1 Year	t+2 Years	t	t+1 Year	t+2 Years	t	t+1 Year	t+2 Years
8603	—	—	5.0	—	—	5.5	—	—	5.4
8604	—	—	14.0	—	—	5.6	—	—	5.5
8610	—	—	7.5	—	—	5.6	—	—	5.5
8704	—	6.8	6.1	—	5.6	5.7	—	5.5	5.5
8809	4.7	4.8	6.3	5.6	5.7	6.0	5.5	5.5	5.9
8810	4.9	5.0	5.6	5.6	5.7	6.0	5.5	5.6	5.9
8903	5.2	4.3	4.8	5.7	5.9	6.1	5.6	5.7	5.9
8909	5.1	5.7	6.4	5.7	6.0	6.4	5.5	5.9	6.3
8910	4.9	5.3	6.1	5.7	6.0	6.5	5.6	5.9	6.5
9003	6.6	7.2	8.4	5.9	6.1	6.6	5.7	5.9	6.5
9009	17.2	17.6	17.9	6.0	6.4	6.4	5.9	6.3	6.2
9010	5.0	6.7	7.7	6.0	6.5	6.4	5.9	6.5	6.2
9104	5.6	6.3	6.5	6.2	6.5	6.4	6.1	6.4	6.2
9203	5.8	5.4	5.7	6.6	6.4	6.4	6.5	6.2	6.2
9210	5.4	6.3	6.2	6.4	6.5	6.5	6.2	6.3	6.3
9304	4.8	5.5	5.9	6.4	6.5	6.6	6.2	6.2	6.3
9310	5.1	6.0	5.6	6.5	6.5	6.6	6.3	6.3	6.3
9404	5.5	6.5	7.1	6.5	6.6	6.5	6.2	6.3	6.6
9904	5.1	6.0	—	6.6	6.5	—	6.3	6.6	—
9604	7.4	—	—	6.5	—	—	6.6	—	—
Total	5.4	5.9	6.5	6.1	6.2	6.0	5.9	6.0	5.9

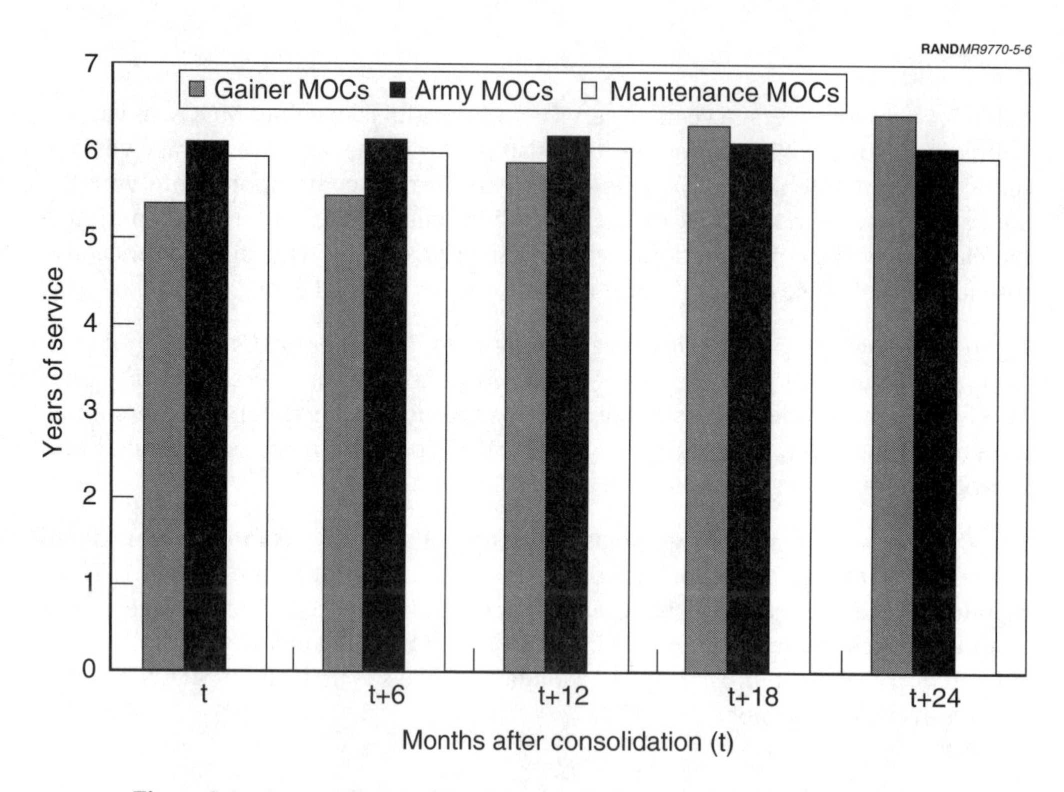

Figure 5.6—Average Years of Service at and After Consolidation for Gainer, Army, and Maintenance MOCs

Table 5.5

Average Grade in Gainer, Army Enlisted, and Maintenance MOCs at Selected Periods

	Gainer MOCs			Army MOCs			Maintenance MOCs		
Date (t)	t	t+1 Year	t+2 Years	t	t+1 Year	t+2 Years	t	t+1 Year	t+2 Years
8603	—	—	4.2	—	—	4.3	—	—	4.3
8604	—	—	6.3	—	—	4.4	—	—	4.3
8610	—	—	4.7	—	—	4.3	—	—	4.3
8704	—	4.6	4.2	—	4.4	4.3	—	4.3	4.3
8809	4.0	4.0	4.5	4.3	4.3	4.4	4.3	4.2	4.3
8810	4.1	4.0	4.2	4.3	4.3	4.4	4.3	4.2	4.3
8903	4.1	3.8	4.0	4.3	4.4	4.4	4.3	4.3	4.3
8909	4.2	4.4	4.6	4.3	4.4	4.5	4.2	4.3	4.4
8910	4.1	4.2	4.4	4.3	4.4	4.5	4.2	4.3	4.5
9003	4.4	4.6	5.0	4.4	4.4	4.5	4.3	4.3	4.5
9009	7.0	7.1	7.2	4.4	4.5	4.4	4.3	4.4	4.4
9010	4.0	4.5	4.9	4.4	4.5	4.5	4.3	4.5	4.4
9104	4.2	4.5	4.6	4.5	4.5	4.5	4.4	4.4	4.5
9203	4.2	4.1	4.2	4.5	4.5	4.5	4.5	4.4	4.4
9210	4.1	4.4	4.2	4.5	4.5	4.4	4.4	4.4	4.3
9304	3.9	4.1	4.2	4.5	4.5	4.5	4.5	4.4	4.4
9310	4.3	4.5	4.5	4.5	4.4	4.5	4.4	4.3	4.4
9404	3.9	4.4	4.7	4.5	4.5	4.5	4.4	4.4	4.5
9904	3.9	4.4	—	4.5	4.5	—	4.4	4.5	—
9604	4.6	—	—	4.5	—	—	4.5	—	—
Total	4.1	4.3	4.5	4.4	4.4	4.4	4.3	4.4	4.4

As shown in Figure 5.7, on average compared to the Army and all maintenance MOCs, gainer MOCs are somewhat more highly graded 24 months after consolidation.

Finally, Table 5.6 details average time in grade in months for each gainer MOC at various points during the consolidation period.

The temptation, with Tables 5.4 through 5.6, is to view an overall increase in experience level as a positive effect of all consolidations. However, during the drawdown period, the tendency was to keep more senior personnel while reducing recruiting, yielding an older, more experienced force overall, but not necessarily in each particular MOC. However, two years after consolidation, the personnel who were assigned to gainer MOCs were at least as senior as their counterparts elsewhere in the Army and in maintenance fields.

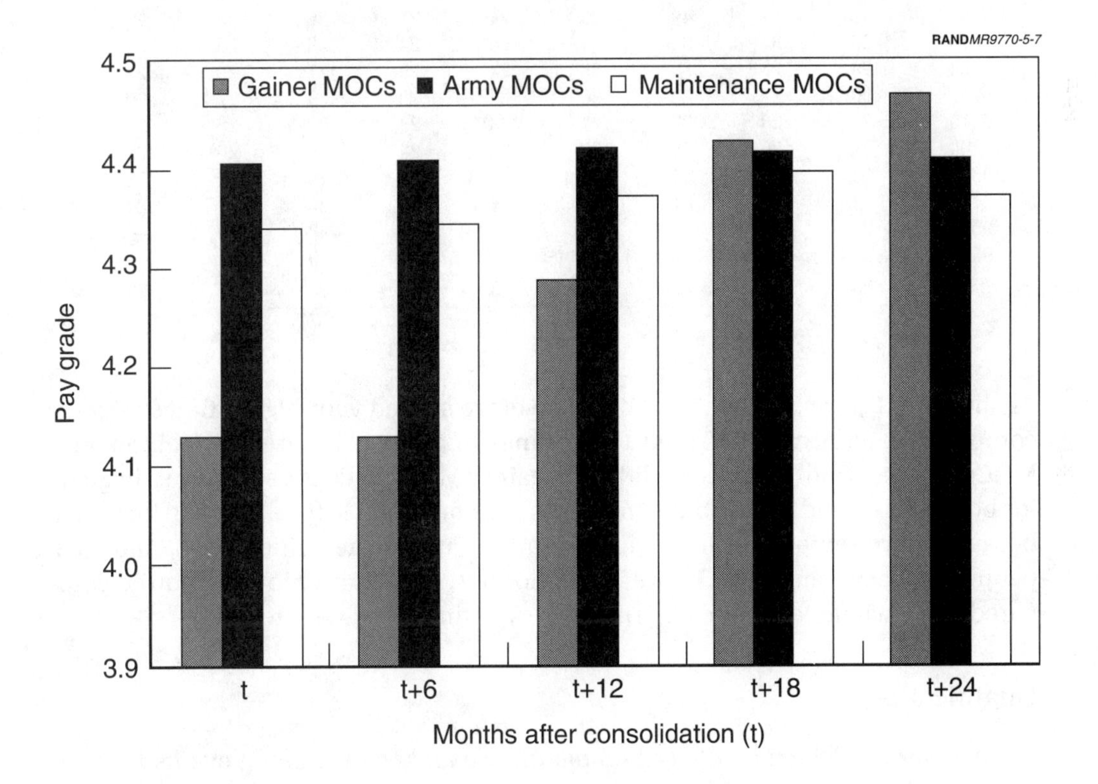

Figure 5.7—Average Grade at and After Consolidation for Gainer, Army, and Maintenance MOCs

Table 5.6

Average Time in Grade in Gainer, Army Enlisted, and Maintenance MOCs at Selected Periods (months)

	Gainer MOCs			Army MOCs			Maintenance MOCs		
Date (t)	t	t+1 Year	t+2 Years	t	t+1 Year	t+2 Years	t	t+1 Year	t+2 Years
8603	—	—	26	—	—	20	—	—	23
8604	—	—	47	—	—	21	—	—	24
8610	—	—	30	—	—	22	—	—	25
8704	—	30	24	—	21	22	—	24	25
8809	20	23	30	21	21	23	24	25	26
8810	18	21	22	22	23	24	25	25	26
8903	26	22	25	22	23	25	26	26	26
8909	21	24	26	21	23	24	25	26	28
8910	21	24	28	23	24	26	25	26	29
9003	30	33	35	23	25	26	26	26	29
9009	49	51	49	23	24	22	26	28	27
9010	28	34	34	24	26	24	26	29	27
9104	23	26	28	25	24	23	27	29	26
9203	30	27	28	26	23	23	29	27	26
9210	27	28	26	24	23	23	27	26	27
9304	23	26	28	23	22	24	26	26	28
9310	23	29	24	23	23	27	26	27	28
9404	19	26	31	22	24	25	26	28	29
9904	20	26	—	24	25	—	28	29	—
9604	33	—	—	25	—	—	29	—	—
Total	23	26	28	23	23	23	26	27	26

As shown in Figure 5.8, gainer MOCs on average started with higher times in grade compared to all Army MOCs but lower times in grade compared to maintenance MOCs. By 24 months after consolidation, gainer MOCs had exceeded times in grade for both. This could be attributed to at least two factors: (1) time in grade increased because more senior people at higher grades entered the gainer MOC, and such people typically wait longer between promotions, and (2) promotions slowed compared to all maintenance and all Army MOCs, so time in grade extended.

Qualification

For personnel affected by the consolidation, we looked at training events from six months prior to 2 years subsequent to a consolidation. We excluded those training courses that were not MOC specific, e.g., "Primary Leadership." Data for training events that occurred prior to 1993 were not broken out by course title or course length. We knew only that a person took a course, not the kind or duration. The corresponding course-length entries, in Table 5.7, list *Unknown*. Table 5.7 reports the average duration, in days, of training courses taken and the percentage of affected personnel who took the courses. These data show that few took formal courses, suggesting that a considerable amount of on-the-job training was given, that no training was required, or that individuals failed to get needed training.

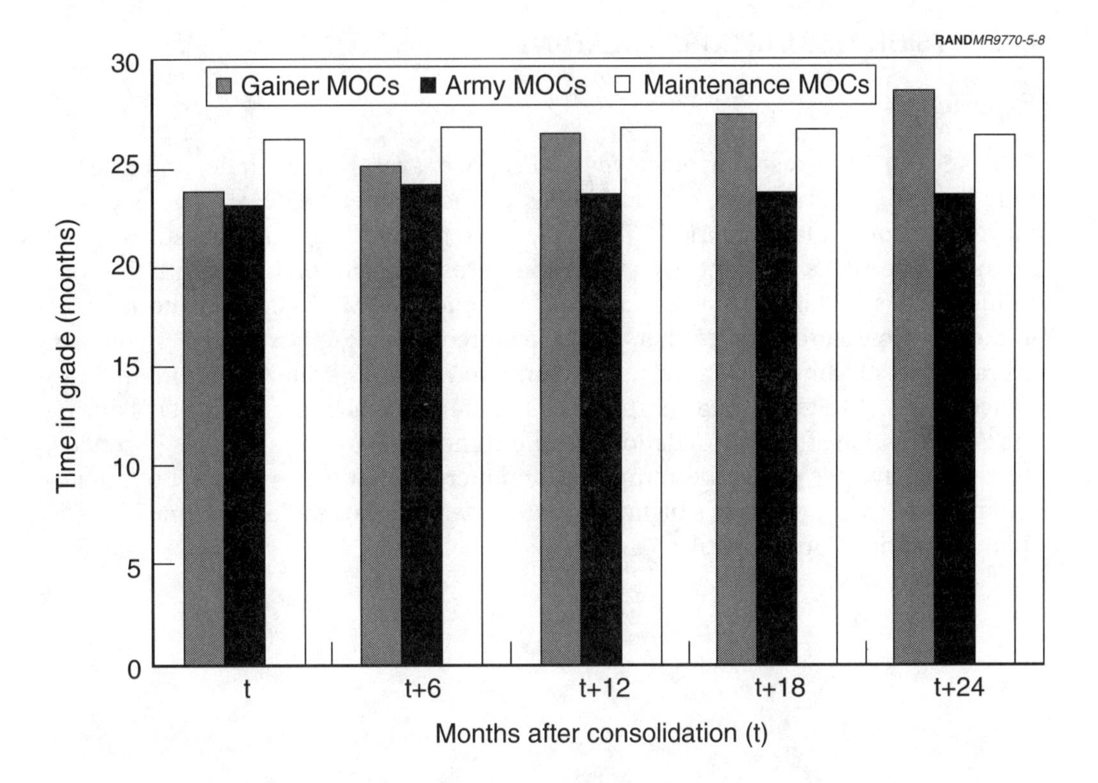

Figure 5.8—Average Time in Grade at and After Consolidation for Gainer, Army, and Maintenance MOCs

Table 5.7

Affected Personnel Taking Formal Training Course and Average Duration

Date of Consoli-dation	Number in Losing MOC	Number Taking Formal Course	Percentage Taking Formal Course	Average Days
8809	1,964	330	17	Unknown
8810	2,611	375	14	Unknown
8903	2,516	287	11	Unknown
8909	1,268	207	16	Unknown
8910	231	49	21	Unknown
9003	2	1	50	Unknown
9009	23	2	9	Unknown
9010	1,993	354	18	Unknown
9104	1,186	143	12	5
9203	49	21	14	32
9210	1,454	143	10	18
9310	9,475	732	8	69
9404	3,583	680	19	100
9504	18,368	1,767	10	98
9604	1,097	40	4	109

EFFECTS ON MARINE CORPS READINESS

Experience

Tables 5.8 to 5.10 present average years of service, average pay grade, and average time in grade, respectively, for individuals in each gainer MOC, at various points during the consolidation period. The tables also present the same measures for all maintenance MOCs and for the total Marine Corps, contemporaneous with the consolidation. As the tables show, some of the consolidating MOCs became more senior in the three measures, while others became more junior. Figures 5.9 to 5.10 present averages for all gainer MOCs over time compared to the Marine Corps and maintenance MOCs. As a group and compared to maintenance MOCs, gainer MOCs during the 2-year period after consolidation are about the same in average years of service, decrease in average pay grade compared, and increase in time in grade. Both maintenance and gainer MOCs are higher in years of service, pay grade, and time in grade than the Marine Corps overall.

Table 5.8

Average Years of Service in Gainer, Marine Enlisted, and Maintenance MOCs at Selected Times

		Gainer MOCs			Marine MOCs			Maintenance MOCs		
Gainer	Date (t)	t	t+12 Year	t+24 Years	t	t+12 Year	t+24 Years	t	t+12 Year	t+24 Years
6073	8807	4.0	5.5	6.0	4.7	4.8	4.9	5.9	5.9	6.0
5977	8908	11.6	12.2	13.3	4.8	4.9	5.1	5.9	6.0	6.2
6312	8908	7.1	7.2	8.7	4.8	4.9	5.1	5.9	6.0	6.2
6313	8908	5.5	5.7	5.5	4.8	4.9	5.1	5.9	6.0	6.2
6314	8908	7.1	7.9	9.8	4.8	4.9	5.1	5.9	6.0	6.2
6315	8908	6.7	5.9	5.9	4.8	4.9	5.1	5.9	6.0	6.2
6317	8908	5.6	6.0	5.7	4.8	4.9	5.1	5.9	6.0	6.2
6318	8908	10.3	5.9	5.0	4.8	4.9	5.1	5.9	6.0	6.2
6322	8908	5.9	6.7	6.8	4.8	4.9	5.1	5.9	6.0	6.2
6323	8908	5.3	5.6	5.7	4.8	4.9	5.1	5.9	6.0	6.2
6324	8908	6.5	5.8	5.5	4.8	4.9	5.1	5.9	6.0	6.2
6337	8908	7.2	6.5	6.6	4.8	4.9	5.1	5.9	6.0	6.2
6353	8908	6.5	6.9	6.2	4.8	4.9	5.1	5.9	6.0	6.2
6363	8908	6.2	7.7	7.7	4.8	4.9	5.1	5.9	6.0	6.2
5957	9404	—	19.0	—	5.0	4.9	—	6.4	6.4	—
4616	9504	—	—	—	4.9	—	—	6.4	—	—

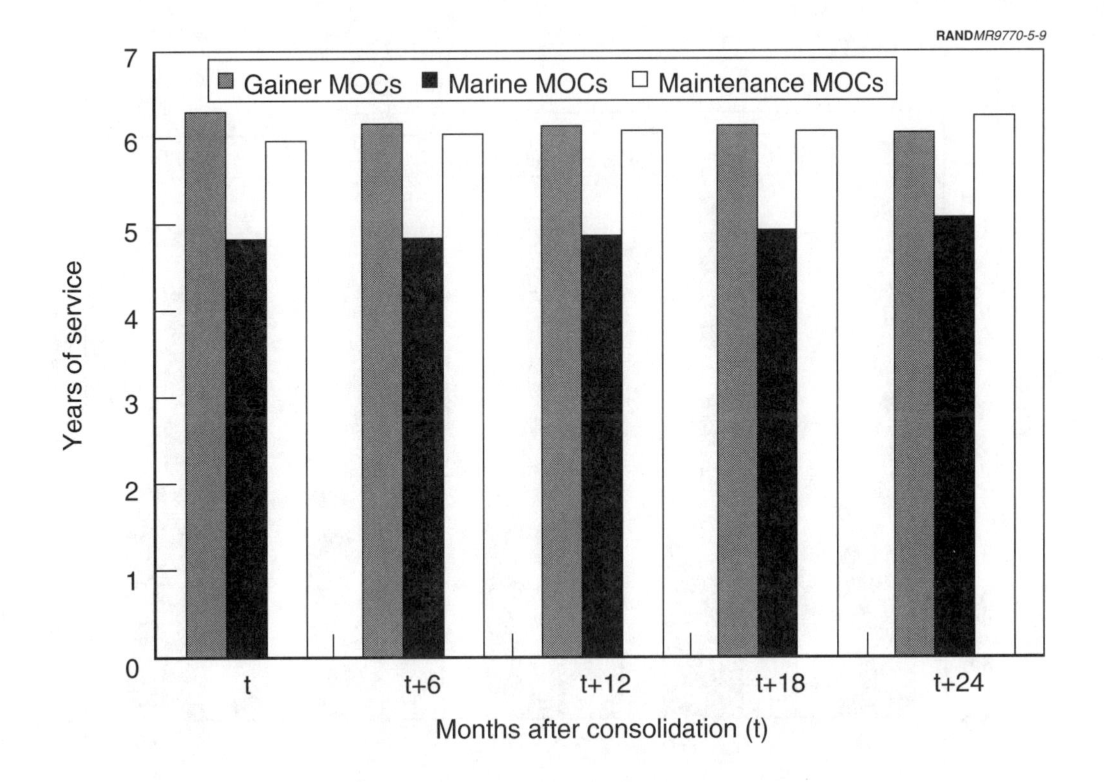

Figure 5.9—Average Years of Service at and After Consolidation for Gainer, Marine, and Maintenance MOCs

Table 5.9

Average Grade in Gainer, Marine Enlisted , and Maintenance MOCs at Selected Periods

		Gainer MOCs			Marine MOCs			Maintenance MOCs		
Gainer	Date (t)	t	t+12 Year	t+24 Years	t	t+12 Year	t+24 Years	t	t+12 Year	t+24 Years
6073	8807	3.8	4.3	4.4	3.9	3.9	3.9	4.4	4.4	4.4
5977	8908	5.9	6.0	6.1	3.9	3.9	3.9	4.4	4.4	4.4
6312	8908	4.6	4.5	4.9	3.9	3.9	3.9	4.4	4.4	4.4
6313	8908	4.2	4.1	4.0	3.9	3.9	3.9	4.4	4.4	4.4
6314	8908	4.6	4.7	5.2	3.9	3.9	3.9	4.4	4.4	4.4
6315	8908	4.6	4.3	4.2	3.9	3.9	3.9	4.4	4.4	4.4
6317	8908	4.4	4.3	4.2	3.9	3.9	3.9	4.4	4.4	4.4
6318	8908	5.3	4.3	4.0	3.9	3.9	3.9	4.4	4.4	4.4
6322	8908	4.5	4.6	4.5	3.9	3.9	3.9	4.4	4.4	4.4
6323	8908	4.3	4.3	4.2	3.9	3.9	3.9	4.4	4.4	4.4
6324	8908	4.8	4.4	4.1	3.9	3.9	3.9	4.4	4.4	4.4
6337	8908	4.8	4.4	4.3	3.9	3.9	3.9	4.4	4.4	4.4
6353	8908	4.5	4.5	4.2	3.9	3.9	3.9	4.4	4.4	4.4
6363	8908	4.4	4.8	4.6	3.9	3.9	3.9	4.4	4.4	4.4
5957	9404	—	7.0	—	3.9	3.9	—	4.4	4.4	—
4616	9504	—	—	—	3.9	—	—	4.4	—	—

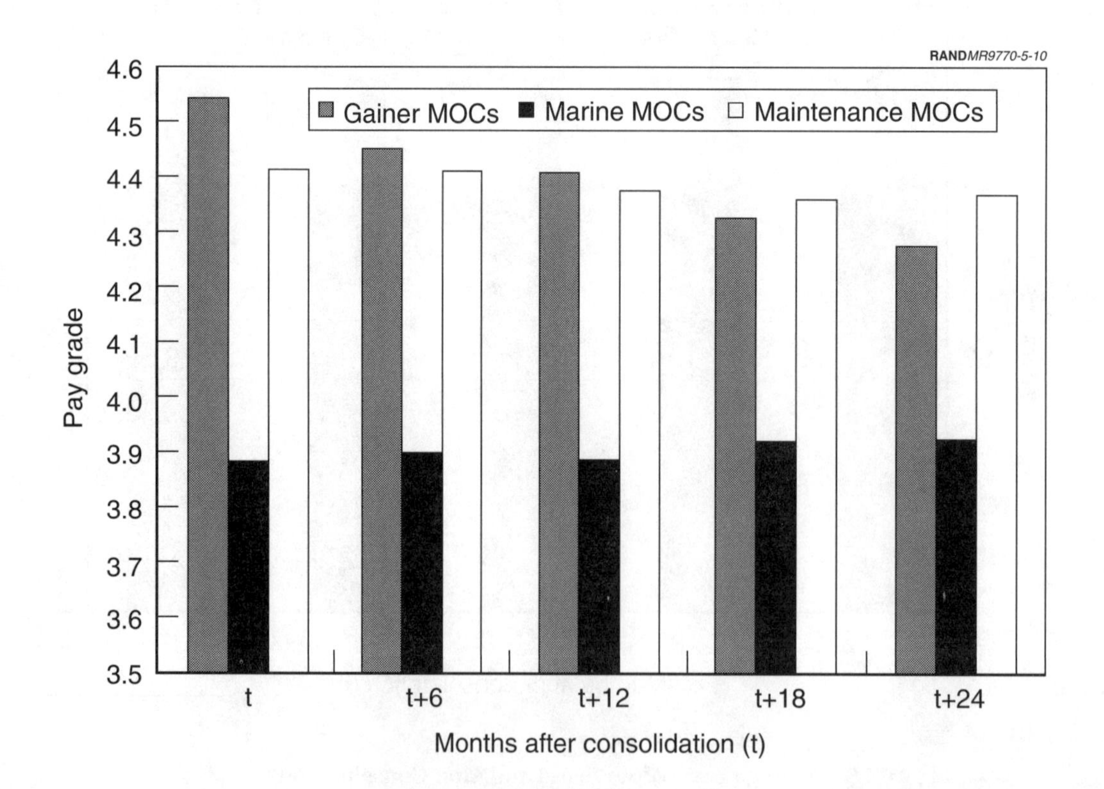

Figure 5.10—Average Pay Grade at and After Consolidation for Gainer, Marine, and Maintenance MOCs

Table 5.10

Average Time in Grade in Gainer, Marine Enlisted, and Maintenance MOCs at Selected Periods (months)

		Gainer MOCs			Marine MOCs			Maintenance MOCs		
Gainer	Date (t)	t	t+12 Year	t+24 Years	t	t+12 Year	t+24 Years	t	t+12 Year	t+24 Years
6073	8807	20	24	25	16	16	16	23	23	24
5977	8908	51	58	64	16	16	17	23	24	25
6312	8908	34	37	40	16	16	17	23	24	25
6313	8908	28	26	26	16	16	17	23	24	25
6314	8908	34	34	44	16	16	17	23	24	25
6315	8908	27	26	29	16	16	17	23	24	25
6317	8908	28	34	28	16	16	17	23	24	25
6318	8908	74	34	27	16	16	17	23	24	25
6322	8908	25	31	35	16	16	17	23	24	25
6323	8908	20	24	29	16	16	17	23	24	25
6324	8908	23	25	28	16	16	17	23	24	25
6337	8908	35	35	35	16	16	17	23	24	25
6353	8908	32	37	36	16	16	17	23	24	25
6363	8908	24	28	37	16	16	17	23	24	25
5957	9404	—	63	—	14	14	—	22	22	—
4616	9504	—	—	—	14	—	—	22	—	—

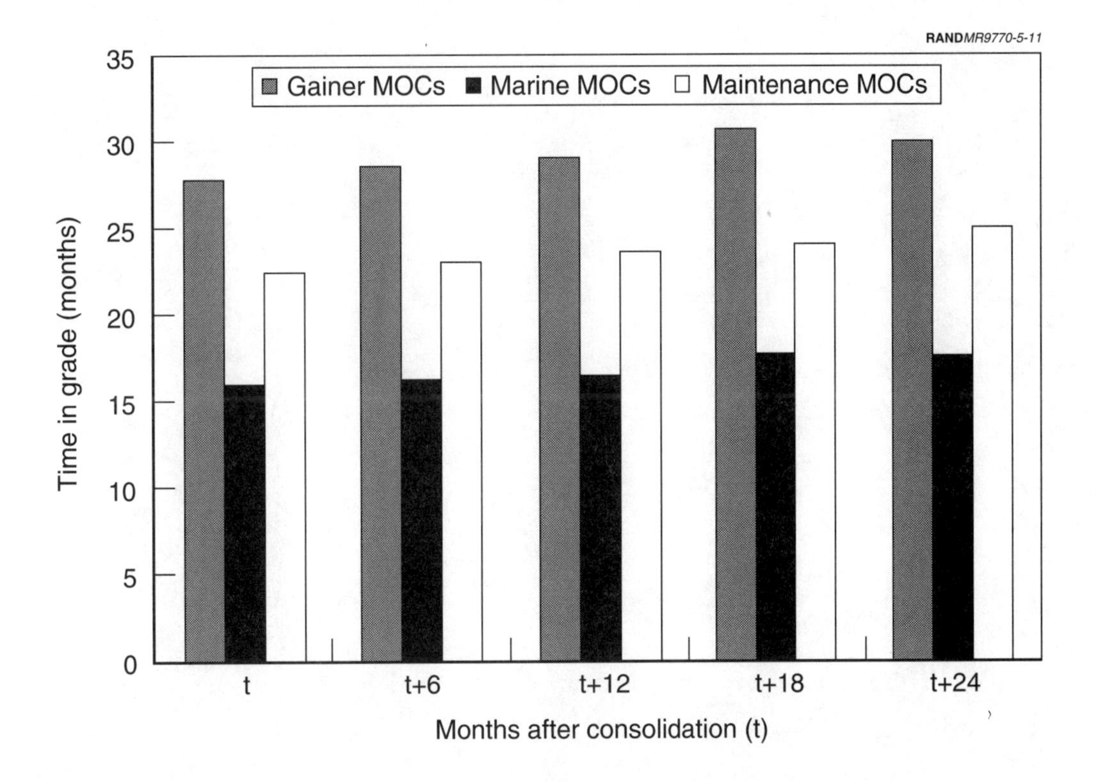

Figure 5.11—Average Time in Grade at and After Consolidation for Gainer, Marine, and Maintenance MOCs

Chapter Six

CONCLUSIONS

The Army, Air Force, and Marine Corps have seen a reduction in the number of MOCs, beginning even before the strength drawdown. Indeed, MOCs have seen a great deal of activity—duty transfers, obsolescence, and consolidation—that make the years examined a dynamic period.

Both the Army and the Marine Corps have detailed and carefully considered processes for evaluating and implementing changes in occupational classifications. A serious effort is made at the beginning of the process to determine whether a proposed change fits the mission of the organization, existing structures, and other policies. Both Services consider the big picture. If nothing else, the sheer volume of work and coordination required to make a change demands careful thought and planning. Required initial analysis and data collection and the involvement of a variety of agencies and experts helps guard against narrow or poorly informed decisions.

Implementation of MOC changes also follows well-documented and well-planned procedures. Responsibilities are clearly laid out, tasks are assigned, and routine communication is established. As a result, when occupational skills are consolidated, training courses are developed, affected agencies are contacted, unit commanders and affected individuals are notified, and individuals are tracked to see that they receive training and are reclassified appropriately. The systems used in both Services make continuous adjustments. If a particular skill consolidation or elimination turns out in practice to cause serious problems, it is likely that the problems will be noticed and that further modifications to the MOC will be made. This has happened in the past in both the Army and the Marine Corps.

Change in an organization is always disruptive, particularly in large, complex organizations. Elimination and consolidation of skill specialties has in the past and will in the future cause some problems and friction, even if the change is necessary and desirable. The procedures in place in the Army and the Marine Corps appear to be designed both to prevent adoption of undesirable changes and to minimize friction when change is needed.

Within the context of our readiness framework, we found no deleterious effects of maintenance MOC consolidations on readiness. The *experience* measures showed that the consolidated MOC became more senior and experienced after two years than all maintenance and Army-wide MOCs. In terms of *availability*, the consolidated MOC had higher percentage of fill than its counterparts. The *qualification* measure is not as clear, suggesting that some portion of training is done on the job, not done at all, or is not needed. This may be disruptive at the unit level, but we had no way of quantifying the level of disturbance, if it exists.

While we did not examine the Marines at the same level of detail as the Army, the *experience* measures indicated about the same results. In general, experience level in the consolidated MOCs was the same as that in all maintenance MOCs and the Marine Corps as a whole; the consolidated MOCs had lower average grades but higher time in grade than maintenance MOCs as a whole.

We recognize that our readiness framework addresses aggregate measures. Furthermore, we were able to examine the data over a two-year transition period, which allowed the effects of temporary disruptions to smooth out. For individuals and for units, the processes of enlarging, eliminating, and consolidating jobs on a day-to-day basis would seem less smooth than do aggregate data over longer periods of time because such change is not immediate and because of frictions in personnel management processes during the period of transition.

Appendix A

ARMY MAINTENANCE MOC CONSOLIDATIONS

Appendix A. Army Maintenance MOC Consolidations

Gainer	Loser	Date	Gainer Title	Loser Title
24H	24S	198910	HAWK Fire Control Repairer	Roland System Mechanic
24H	24V	198810	HAWK Fire Control Repairer	HAWK Maintenance Chief
24K	24S	198910	HAWK Continuous Wave Radar Repairer	Roland System Mechanic
24K	24V	198810	HAWK Continuous Wave Radar Repairer	HAWK Maintenance Chief
25L	26H	199003	Air Defense Artillery Command and Control System Operator/Repairer	Air Defense Radar Repairer
26C	26B	198410	Target Acquisition Surveillance Radar Repairer	Weapons Support Radar Repairer
26L	26H	199003	Tactical Microwave Systems Repairer	Air Defense Radar Repairer
26V	26R	198510	Strategic Microwave Systems Repairer	Strategic Microwave Systems Operator
27E	27B	199504	Land Combat Electronic Missile System Repairer	Land Combat Support System (LCSS) Test Specialist
27E	27H	198406	TOW/Dragon Missile Electronics Repairer	Shillelagh Repairer
27H	24L	198810	HAWK Firing Section Repairer	HAWK Launcher And Mechanical Systems Repairer
27H	24V	198810	HAWK Firing Section Repairer	HAWK Maintenance Chief
27H	27J	199504	HAWK Firing Section Repairer	HAWK Field Maintenance Equipment/Pulse Acquisition Radar Repairer
27J	24J	198809	HAWK Field Maintenance Equipment/Pulse Acquisition Radar Repairer	HAWK Pulse Radar Repairer
27J	24S	198809	HAWK Field Maintenance Equipment/Pulse Acquisition Radar Repairer	Roland System Mechanic
27J	24V	198809	HAWK Field Maintenance Equipment/Pulse Acquisition Radar Repairer	HAWK Maintenance Chief
27M	27B	199504	Multiple Launch Rocket System (MLRS) Repairer	Land Combat Support System (LCSS) Test Specialist
27V	24S	198809	HAWK Maintenance Chief	Roland System Mechanic
27V	24V	198809	HAWK Maintenance Chief	HAWK Maintenance Chief
27Z	27V	199404	Land Combat/Air Defense Systems Maintenance Chief	HAWK Maintenance Chief
29E	31E	198603	Radio Repairer	Field Radio Repairer
29E	32H	198603	Radio Repairer	Fixed Station Radio Repairer
29F	32F	198509	Fixed Communications Security Equipment Repairer	Fixed Ciphony Repairer
29F	32G	198504	Fixed Communications Security Equipment Repairer	Fixed Cryptographic Equipment Repairer
29G	34F	198610	Digital Communications Equipment Repairer	Digital Subscriber Terminal Equipment (DSTE) Repairer
29H	34H	198610	Automatic Digital Message Switch Equipment Repairer	Automatic Digital Message Switch Equipment (ADMSE) Repairer

Gainer	Loser	Date	Gainer Title	Loser Title
29J	29J	198603	Telecommunications Terminal Device Repairer	Telecommunications Terminal Device Repairer
29J	31J	198510	Telecommunications Terminal Device Repairer	Teletypewriter Repairer
29J	39K	198903	Telecommunications Terminal Device Repairer	IBM Automatic Data Processing Systems (ADPS) Repairer
29J	39L	199203	Telecommunications Terminal Device Repairer	Field Artillery Digital Systems Repairer
29J	39T	198910	Telecommunications Terminal Device Repairer	Tactical Computer Systems Repairer
29J	39Y	199203	Telecommunications Terminal Device Repairer	Field Artillery Tactical Fire Direction Systems Repairer
29M	26L	198603	Tactical Satellite/Microwave Repairer	Tactical Microwave Systems Repairer
29N	36H	198509	Switching Central Repairer	Dial/Manual Central Office Repairer
29P	31S	198509	Communications Security Maintenance Chief	Field General Communication Security Repairer
29S	29F	199010	Communications Security Equipment Repairer	Fixed Communications Security Equipment Repairer
29S	29P	199010	Communications Security Equipment Repairer	Communications Security Maintenance Chief
29S	31S	198509	Communications Security Equipment Repairer	Field General Communication Security Repairer
29S	31T	198509	Communications Security Equipment Repairer	Field Systems Communications Security Equipment Repairer
29U	34F	198610	Digital Equipment Maintenance Chief	Digital Subscriber Terminal Equipment (DSTE) Repairer
29U	34H	198610	Digital Equipment Maintenance Chief	Automatic Digital Message Switch Equipment (ADMSE) Repairer
29V	26V	198610	Microwave Systems Operator/Repairer	Strategic Microwave Systems Repairer
29V	29M	199104	Microwave Systems Operator/Repairer	Tactical Satellite/Microwave Repairer
29V	29T	199104	Microwave Systems Operator/Repairer	Satellite Microwave Communications Chief
29W	32Z	198604	Communications Electronics Maintenance Chief	Communications-Electronics Maintenance Chief
29W	39W	199009	Communications Electronics Maintenance Chief	Radar/Special Electronic Devices Maintenance Chief
29X	32Z	198610	Communications Equipment Maintenance Chief	Communications-Electronics Maintenance Chief
29Y	26Y	198610	Satellite Communications (SATCOM) Systems Repairer	Satellite Communications Equipment Repairer
29Y	29T	199104	Satellite Communications (SATCOM) Systems Repairer	Satellite Microwave Communications Chief
29Z	29X	199010	Electronics Maintenance Chief	Communications Equipment Maintenance Chief
29Z	32Z	198610	Electronics Maintenance Chief	Communications-Electronics Maintenance Chief
29Z	34Z	198610	Electronics Maintenance Chief	ADP Maintenance Supervisor
29Z	39X	199009	Electronics Maintenance Chief	Electronics Equipment Maintenance Chief
31F	36L	199504	Network Switching Systems Operator/Maintainer	Transportable Automatic Switching Systems Operator/Maintainer
31F	36M	199210	Electronic Switching Systems Operator	Switching Systems Operator
31L	36C	198610	Wire Systems Installer	Wire Systems Installer
31N	31X	198809	Communications Systems/Circuit Controller	Communications Systems/Circuit Control Supervisor
31N	32D	198809	Communications Systems/Circuit Controller	Communications System Circuit Controller
31P	29V	199404	Microwave Systems Operator-Maintainer	Microwave Systems Operator/Repairer
31P	31N	199404	Microwave Systems Operator-Maintainer	Communications Systems/Circuit Controller
31R	31D	199504	Multichannel Transmission Systems Operator-Maintainer	Mobile Subscriber Equipment (MSE) Transmission System Operator

Gainer	Loser	Date	Gainer Title	Loser Title
31R	31M	199504	Multichannel Transmission Systems Operator-Maintainer	Multichannel Transmission Systems Operator
31S	29Y	199404	Satellite Communications Systems Operator-Maintainer	Satellite Communications (SATCOM) Systems Repairer
31S	31M	199504	Satellite Communications Systems Operator-Maintainer	Multichannel Transmission Systems Operator
31S	31Y	199504	Satellite Communications Systems Operator-Maintainer	Telecommunications Systems Supervisor
31U	31G	199304	Signal Support Systems Specialist	Tactical Communications Chief
31U	31K	199304	Signal Support Systems Specialist	Combat Signaler
31U	31V	199304	Signal Support Systems Specialist	Unit Level Communications Maintainer
31Y	36C	198610	Telecommunications Systems Supervisor	Wire Systems Installer
31Z	29Z	199504	Senior Signal Sergeant	Electronics Maintenance Chief
33M	33S	198406	Electronic Warfare/Intercept Strategic C/C Repairer	Electronic Warfare/Intercept Systems Repairer
33P	33S	198406	Electronic Warfare/Intercept Strategic Receiver Equipment Repairer	Electronic Warfare/Intercept Systems Repairer
33Q	33S	198406	Electronic Warfare/Intercept P/S Strategic Equipment Repairer	Electronic Warfare/Intercept Systems Repairer
33R	33S	198406	Electronic Warfare/Intercept Aviation Systems Repairer	Electronic Warfare/Intercept Systems Repairer
33R	33V	199310	Electronic Warfare/Intercept Aviation Systems Repairer	Electronic Warfare/Intercept Aerial Sensor Repairer
33T	33S	198406	Electronic Warfare/Intercept Tactical Systems Repairer	Electronic Warfare/Intercept Systems Repairer
33V	26E	198604	Electronic Warfare/Intercept Aerial Sensor Repairer	Aerial Radar Sensor Repairer
33V	26F	198604	Electronic Warfare/Intercept Aerial Sensor Repairer	Aerial Photoactive Sensor Repairer
33Y	33M	199010	Strategic Systems Repairer	Electronic Warfare/Intercept Strategic C/C Repairer
33Y	33P	199010	Strategic Systems Repairer	Electronic Warfare/Intercept Strategic Receiver Equipment Repairer
33Y	33Q	199010	Strategic Systems Repairer	Electronic Warfare/Intercept P/S Strategic Equipment Repairer
33Z	33S	198406	Electronic Warfare/Intercept Systems Maintenance Supervisor	Electronic Warfare/Intercept Systems Repairer
35B	27B	199504	Land Combat Support System (LCSS) Test Specialist	Land Combat Support System (LCSS) Test Specialist
35C	39C	199504	Surveillance Radar Repairer	Target Acquisition/Surveillance Radar Repairer
35D	93D	199604	Meteorological Equipment Repairman	Air Traffic Control Equipment Repairer
35E	29E	199504	Radio and Communications Security (COMSEC) Repairer	Radio Repairer
35E	29S	199504	Radio and Communications Security (COMSEC) Repairer	Communications Security Equipment Repairer
35F	39E	199504	Special Electronic Devices Repairer	Special Electronic Devices Repairer
35J	29J	199504	Telecommunications Terminal Device Repairer	Telecommunications Terminal Device Repairer
35L	68L	199604	Avionic Communications Equipment Repairer	Avionic Communications Equipment Repairer
35M	13R	199504	Radar Repairer	Field Artillery (FA) Firefinder Radar Operator
35M	39C	199504	Radar Repairer	Target Acquisition/Surveillance Radar Repairer
35N	29N	199504	Wire Systems Equipment Repairer	Switching Central Repairer
35P	26K	198604	Avionic Equipment Maintenance Supervisor	Aerial Early Warning/Defense Equipment Repairer
35Q	68Q	199604	Avionic Flight Systems Repairer	Avionic Flight Systems Repairer
35R	68R	199604	Avionic Special Equipment Repairer	Avionic Radar Repairer

Gainer	Loser	Date	Gainer Title	Loser Title
35W	29S	199504	Electronic Maintenance Chief	Communications Security Equipment Repairer
35W	29W	199504	Electronic Maintenance Chief	Communications Electronics Maintenance Chief
35W	29Z	199504	Electronic Maintenance Chief	Electronics Maintenance Chief
35Y	27B	199504	Integrated Family of Test Equipment Operator/Maintainer	Land Combat Support System (LCSS) Test Specialist
35Z	29Z	199504	Senior Electronics Maintenance Chief	Electronics Maintenance Chief
35Z	93D	199604	Senior Electronics Maintenance Chief	Air Traffic Control Equipment Repairer
39B	35C	198603	Automatic Test Equipment Operator/Maintainer	Automatic Test Equipment (ATE) Repairer
39C	26C	198610	Target Acquisition/Surveillance Radar Repairer	Target Acquisition Surveillance Radar Repairer
39D	34C	198610	DAS3 Computer Systems Repairer	Decentralized Automated Service Support System
39E	35E	198704	Special Electronic Devices Repairer	Special Electronic Devices Repairer
39G	29G	198903	Automated Communications Computer Systems Repairer	Digital Communications Equipment Repairer
39G	29H	198903	Automated Communications Computer Systems Repairer	Automatic Digital Message Switch Equipment Repairer
39G	39D	199404	Automated Communications Computer Systems Repairer	Decentralized Automated Service Support System (DAS3) Computer Systems Repairer
39G	39V	199404	Automated Communications Computer Systems Repairer	Computerized Systems Maintenance Chief
39K	34J	198410	IBM Automatic Data Processing Systems (ADPS) Repairer	Univac 1004/1005 DCT 9000 System Repairer
39K	34K	198410	IBM Automatic Data Processing Systems (ADPS) Repairer	IBM 3000 Repairer
39L	34L	198704	Field Artillery Digital Systems Repairer	Field Artillery Digital Systems Repairer
39T	34T	198704	Tactical Computer Systems Repairer	Tactical Computer Systems Repairer
39V	29U	198810	Computerized Systems Maintenance Chief	Digital Equipment Maintenance Chief
39V	34C	198704	Computerized Systems Maintenance Chief	Decentralized Automated Service Support System
39V	34J	198704	Computerized Systems Maintenance Chief	Univac 1004/1005 DCT 9000 System Repairer
39V	34K	198704	Computerized Systems Maintenance Chief	IBM 3000 Repairer
39X	34Z	198610	Electronics Equipment Maintenance Chief	ADP Maintenance Supervisor
39Y	34Y	198610	Field Artillery Tactical Fire Direction Systems Repairer	Field Artillery Tactical Fire Direction Systems
41B	41E	198909	Topographic Instrument Repair Specialist	Audio-Visual Equipment Repairer
45B	45L	199210	Small Arms/Artillery Repairer	Artillery Repairer
45G	41C	199210	Fire Control Repairer	Fire Control Instrument Repairer
45K	41C	199210	Armament Repairer	Fire Control Instrument Repairer
45K	45L	199210	Armament Repairer	Artillery Repairer
45K	45Z	199210	Armament Repairer	Armament/Fire Control Maintenance Supervisor
55B	55R	199304	Ammunition Specialist	Ammunition Stock Control and Accounting Specialist
55B	55X	199304	Ammunition Specialist	Ammunition Inspector
55B	55Z	199604	Ammunition Specialist	Ammunition Supervisor
67B	67A	199107	Certified General Aircraft Repairer	General Aircraft Repairer

Gainer	Loser	Date	Gainer Title	Loser Title
67G	66G	198909	Utility Airplane Repairer	Utility Airplane Technical Inspector
67H	66H	198909	Observation Airplane Repairer	Observation Airplane Technical Inspector
67N	66N	198909	UH-1 Helicopter Repairer	Utility Helicopter Technical Inspector
67R	66R	198909	AH-64 Attack Helicopter Repairer	AH-64 Attack Helicopter Technical Inspector
67S	66S	198909	OH-58D Helicopter Repairer	Scout Helicopter Technical Inspector
67T	66T	198909	UH-60 Helicopter Repairer	Tactical Transport Helicopter Technical Inspector
67U	66U	198909	CH-47 Helicopter Repairer	Medium Helicopter Technical Inspector
67V	66V	198909	Observation/Scout Helicopter Repairer	Observation/Scout Helicopter Technical Inspector
67X	66X	198909	Heavy Lift Helicopter Repairer	Heavy Lift Helicopter Technical Inspector
67Y	66Y	198909	AH-1 Attack Helicopter Repairer	AH-1 Attack Helicopter Technical Inspector
68J	66J	198910	Aircraft Armament/Missile Systems Repairer	Aircraft Armament Technical Inspector
68J	68M	198810	Aircraft Armament/Missile Systems Repairer	Aircraft Weapon Systems Repairer
68L	35L	198903	Avionic Communications Equipment Repairer	Avionic Communications Equipment Repairer
68L	35P	198903	Avionic Communications Equipment Repairer	Avionic Equipment Maintenance Supervisor
68N	35K	198810	Avionic Mechanic	Avionics Mechanic
68N	35P	198810	Avionic Mechanic	Avionic Equipment Maintenance Supervisor
68P	35P	198810	Avionic Maintenance Supervisor	Avionic Equipment Maintenance Supervisor
68Q	35P	198903	Avionic Flight Systems Repairer	Avionic Equipment Maintenance Supervisor
68R	35P	198903	Avionic Radar Repairer	Avionic Equipment Maintenance Supervisor
68R	35R	198903	Avionic Radar Repairer	Avionic Special Equipment Repairer
74G	36L	199504	Telecommunications Computer Operator-Maintainer	Transportable Automatic Switching Systems Operator/ Maintainer
74G	39G	199504	Telecommunications Computer Operator-Maintainer	Automated Communications Computer Systems Repairer
82B	41B	198610	Construction Surveyor	Topographic Instrument Repair Specialist
82D	41B	198610	Topographic Surveyor	Topographic Instrument Repair Specialist
88L	61C	198610	Watercraft Engineer	Watercraft Engineer
88P	65B	198610	Locomotive Repairer	Locomotive Repairer
88P	88Q	199404	Locomotive Repairer	Railway Car Repairer
88P	88R	199404	Locomotive Repairer	Airbrake Repairer
88P	88S	199404	Locomotive Repairer	Locomotive Electrician
88Q	65D	198610	Railway Car Repairer	Railway Car Repairer
88R	65E	198610	Airbrake Repairer	Airbrake Repairer
88S	65F	198610	Locomotive Electrician	Locomotive Electrician
93C	26D	198510	Air Traffic Control (ATC) Operator	Ground Control Approach Radar Repairer

Appendix B

MARINE CORPS MAINTENANCE MOC CONSOLIDATIONS

Appendix B. Marine Corps Maintenance MOC Consolidations

Gainer	Loser	Date	Gainer Title	Loser Title
6073	6077	198807	Aircraft Maintenance Ground Support Equipment Electrician	Aircraft Maintenance Ground Support Equipment
6073	6078	198807	Aircraft Maintenance Ground Support Equipment Electrician	Aircraft Maintenance GSE Refrigeration Mechanic
5977	6374	198908	Tactical General Purpose Computer Technician	Imagery Interpretation Equipment Repairer
6312	6332	198908	Aircraft Communications/Navigation/Electrical/Weapon/DECM	Aircraft Electronic Systems Technician A-4/TA-4/OA-4
6312	6352	198908	Aircraft Communications/Navigation/Electrical/Weapon/DECM	Aircraft Weapons Systems Specialist A-4/TA-4/OA-4
6313	6365	198908	Aircraft Communications/Navigation/Electrical/Weapon/DECM	Aircraft Communication/Navigation/Radar Systems Technician, EA-6
6314	6334	198908	Aircraft Communications/Navigation/Electrical/Weapon/DECM	Aircraft Electronic Systems Technician RF-4/F-4
6315	6355	198908	Aircraft Communications/Navigation/Electrical/Weapon/DECM	Aircraft Weapons Systems Specialist AV-8
6317	6357	198908	Aircraft Communications/Navigation/Electrical/Weapon/DECM	Aircraft Weapons Systems Specialist F/A-18
6318	6316	198908	Aircraft Communications/Navigation/Electrical/Weapon/DECM	Aircraft Communications/Navigation/Electrical/Weapon/DECM Systems Technician KC-130
6318	6336	198908	Aircraft Communications/Navigation/Electrical/Weapon/DECM	Aircraft Electrical Systems Technician, KC-130
6318	6364	198908	Aircraft Communications/Navigation/Electrical/Weapon/DECM	Aircraft Weapons Systems Specialist, Helicopter
6322	6342	198908	Aircraft Communications/Navigation/Electrical/Weapon/DECM	Aircraft Electronic Systems Technician CH-46
6323	6343	198908	Aircraft Communications/Navigation/Electrical/Weapon/DECM	Aircraft Electronic Systems Technician CH-53
6323	6345	198908	Aircraft Communications/Navigation/Electrical/Weapon/DECM	Aircraft Electronic Systems Technician CH-53E
6324	6344	198908	Aircraft Communications/Navigation/Electrical/Weapon/DECM	Aircraft Electronic Systems Technician U/AH1
6324	6364	198908	Aircraft Communications/Navigation/Electrical/Weapon/DECM	Aircraft Weapons Systems Specialist, Helicopter
6337	6367	198908	Aircraft Electrical Systems Technician, F/A-18	Aircraft Integrated Weapon Systems Technician
6353	6364	198908	Aircraft Weapons Systems Specialist A-6/TC-4C	Aircraft Weapons Systems Specialist, Helicopter
6363	6372	198908	Aircraft Radar Reconnaissance/Camera Systems Technician R	Aerial Camera Systems Technician
5957	5945	199404	Station Air Traffic Control Precision Approach Radar Tech	Aviation Radar Repairer
4616	1542	199504	Reproduction Equipment Repairer	Reproduction Equipment Repairer
5914	5924	199701	Surface to Air Missile Systems Ordnance Technician	Surface Air Defense System Acquisition Technician
5924	5925	199701	Surface Air Defense System Acquisition Technician	Surface Air Defense Systems Fire Control Technician
5925	5928	199701	Surface Air Defense Systems Fire Control Technician	Surface Air Defense Systems Chief

REFERENCES

8th Quadrennial Review of Military Compensation, *Rewarding, Organizing, and Managing People in the 21st Century: Time for a Strategic Approach*, Executive Report, Washington, D.C.: Office of the Assistant Secretary of Defense (Force Management Policy), June 30, 1997.

Bridges, William, *Job Shift: How to Prosper in a Workplace Without Jobs*, Reading, Mass.: Addison-Wesley, 1995.

Department of Defense, Office of the Assistant Secretary of Defense, Personnel and Readiness, *Occupational Conversion Index: Enlisted/Officer/Civilian*, Alexandria, Va.: U.S. Department of Commerce, National Technical Information Center, 1993.

Department of Labor, *High Performance Work Practices and Firm Performance* Washington, D.C., 1993.

Department of Labor, *The Road to High-Performance Workplaces: A Guide to Better Jobs and Better Business Results*, Washington, D.C., 1994.

Department of the Army, "Implementation of Changes to the Military Occupational Classification and Structure," Circular 611-96-2, 1996.

_____, *Military Occupational Classification Structure Development and Implementation*, Washington, D.C.: Headquarters, Department of the Army, Army Regulation 611-1, September 30, 1977.

Donnely, Kate, Peter LeBlanc, Dale Torrence, and Margaret Lyon, "Career Banding," *Human Resource Management*, Vol. 31, No. 1, 1992, pp. 35–43.

Gotz, Glenn A., and Richard E. Stanton, *Modeling the Contribution of Maintenance Manpower to Readiness and Sustainability*, Santa Monica, Calif.: RAND, R-3200-FMP, 1986.

Husilid, Mark A., "The Impact of Human Resource Management Practices on Turnover, Productivity, and Corporate Financial Performance," *Academy of Management Journal*, Vol. 38, No. 3, 1995, pp. 635–672.

Ichniowski, Casey, Kathryn Shaw, and Giovanna Prennushi, "The Effects of Human Resource Management Practices on Productivity," mimeograph, Columbia University, June 10, 1993.

Kirby, Sheila Nataraj, and Harry J. Thie, *Enlisted Personnel Management: A Historical Perspective,* Santa Monica, Calif.: RAND, MR-755-OSD, 1996.

Kirin, Stephen J., and John D. Winkler, *The Army Military Occupational Specialty Database*, Santa Monica, Calif.: RAND, N-3527-A, 1992.

Rice, Donald B., *Defense Resource Management Study*, Washington, D.C.: U.S. Government Printing Office, 1979.

Robbert, Al, Brent Keltner, Ken Reynolds, Mark Spranc, and Beth Benjamin, *Differentiation in Military Human Resource Management,* Santa Monica, Calif.: RAND, MR-838-OSD, 1996.

Schank, John F. et al., *Relating Resources to Personnel Readiness: Use of Army Strength Management Models*, Santa Monica, Calif.: RAND, MR-790-OSD, 1997.

Spence, Floyd D., Chairman, House Committee on National Security, *Military Readiness 1997: Rhetoric and Reality*, Washington, D.C., April 9, 1997.

U.S. General Accounting Office, *Management Practices: U.S. Companies Improve Performance Through Quality Efforts*, Washington, D.C., 1991.

U.S. House of Representatives, Committee on National Security, Report on H.R. 1119, *National Defense Authorization Act for Fiscal Year 1998*, Report 105-132, June 16, 1997.

U.S. Marine Corps, ALMAR 077/96, 1996.

U.S. Marine Corps, Marine Corps Order 1200.15A.

U.S. Marine Corps, Marine Corps Order 5311.1C, draft currently under consideration for final approval.

Wild, Jr., William G., and Bruce R. Orvis, *Design of Field-Based Crosstraining Programs and Implications for Readiness*, 1993.